AF553871

ENTOMOLOGY

Shagufta

APH PUBLISHING CORPORATION
4435-36/7, Ansari Road, Darya Ganj
New Delhi-110 002

Published by

S.B. Nangia
APH Publishing Corporation
4435-36/7, Ansari Road, Darya Ganj
New Delhi-110 002
PH. : 23274050
Email : aphbooks@gmail.com

2021

Rs. 1495/-

Typesetting at

Rawat Computers
Gandhi Nagar, Delhi

Printed at

Balaji Offset
Navin Shahdara, Delhi-32

Contents

Preface *vii*

1. Introduction to Entomology and Its Related Fields 1
   - Entomology
   - Timeline of Entomology
   - Forensic Entomology
   - Forensic Entomological Decomposition
   - Economic Entomology
   - Home Stored Product Entomology

2. Introduction to Insects, Insect Ecology and Other Related Areas 92
   - Insect
   - Insect Ecology
   - Insect Winter Ecology
   - Insectary Plants
   - Insectivore

3. Insect Phylogeny, Physiology, Sterility and Defense in Insects 129
   - Phylogeny
   - Insect Physiology
   - Respiratory System of Insects
   - Eusociality
   - Defense in Insects
   - Inherited Sterility in Insects
   - Sterile Insect Technique

4. Scientific Classification, Subclasses, Orders of Insects and Entomophagy 206
   - Archaeognatha
   - Coxoplectoptera
   - Pterygota

- Entomophagy
- Entomophagous Parasite

5. Insect Biodiversity, Measurement Methods and Insect Migration 227
   - Insect Biodiversity
   - Insect Collecting
   - Insect Trap
   - Insect Migration

6. Other Major Areas of Study of Entomology: A Select Overview 237
   - Entomological Evidence Collection
   - Ethno-entomology
   - Entomological Warfare
   - Medical Entomology
   - List of Entomologists

Bibliography 275

Index 279

Preface

Entomology is the scientific study of insects, a branch of arthropod. At some 1.3 million described species, insects account for more than two-thirds of all known organisms, date back some 400 million years, and have many kinds of interactions with humans and other forms of life on earth. It is a specialty within the field of biology. Though technically incorrect, the definition is sometimes widened to include the study of terrestrial animals in other arthropod groups or other phyla, such as arachnids, myriapods, earthworms, land snails, and slugs. Despite their small size, the sheer numbers or biomass of insects means that they have a significant impact on the environment and therefore upon our lives. Although insects have never adapted to a truly marine environment, they inhabit almost all terrestrial and aquatic environments. Some have adapted to live with us (cockroaches, silverfish) while others live on us (lice, bed bugs, mosquitoes). Their species richness or diversity surpasses any other group of organisms. It has been said that insects outnumber all the other species of animals and plants combined. This relative diversity may be appreciated by a speciescape whereby the size of an organism, as drawn, is approximately proportional to the number of described species in the group (or taxon) that it represents.

We do not know exactly how many species of insects there are. There are about 1 million named species. It has taken about 300 years for scientists to describe these. Estimates of actual species richness vary from less than 5 million to as many as 80 million. Figures at the higher end of these estimates are dependent upon the diversity of species now being recovered from the canopies of rain-forests. Obviously there are many years (if not centuries) of work for insect taxonomists to describe, name and classify these species. This book titled, "Entomology" gives brief introduction to entomology and its related fields. It also provides readers with an introduction to insects, insect ecology and other related areas. Insect phylogeny, physiology, sterility aspects are covered and defense in insects is described. In addition, the scientific classification, subclasses, orders of insects are reflected upon and the subject area of entomophagy is covered. Similarly, insect biodiversity, measurement methods and insect migration are highlighted. Other major areas of entomology are discussed in detail. The book is extremely user friendly as it gives a systematic presentation of existing facts, figures and other details besides bibliography and index.

- Editor

1

Introduction to Entomology and Its Related Fields

ENTOMOLOGY

Entomology is the scientific study of insects, a branch of arthropodology. At some 1.3 million described species, insects account for more than two-thirds of all known organisms, date back some 400 million years, and have many kinds of interactions with humans and other forms of life on earth. It is a specialty within the field of biology. Though technically incorrect, the definition is sometimes widened to include the study of terrestrial animals in other arthropod groups or other phyla, such as arachnids, myriapods, earthworms, land snails, and slugs. Like several of the other fields that are categorized within zoology, entomology is a taxon-based category; any form of scientific study in which there is a focus on insect related inquiries is, by definition, entomology. Entomology therefore includes a cross-section of topics as diverse as molecular genetics, behaviour, biomechanics, biochemistry, systematics, physiology, developmental biology, ecology, morphology, paleontology, anthropology, robotics, agriculture, nutrition, forensic science and more.

History

Entomology is rooted in nearly all human cultures from prehistoric times, primarily in the context of agriculture (especially biological control and beekeeping), but scientific study began only as recently as the 16th century. The list of entomologists through recorded history is enormous.

In Popular Culture

Gil Grissom on the CSI: Crime Scene Investigation TV show is an entomologist, who is played by actor William Petersen. Similarly, entomologist Jack Hodgins of *Bones*, portrayed by TJ Thyne, helps his team by analyzing insects (such as *Hydrotaea*) and "particulates" near to or attached to decomposed victims, often identifying the precise location a murder originally occurred; he is also an expert in botany and mineralogy. In Arthur Conan Doyle's story, *The Hound of the Baskervilles*, the villain is a naturalist who collects butterflies, making him an "evil" entomologist. The Aubrey–Maturin sea novels of Patrick O'Brian have frequent appearances by Sir Joseph Blaine, a Royal Navy intelligence official who is also an avid entomologist. He recruits Dr. Stephen Maturin, one of the principal characters, as a spy. Their conferences on espionage activities invariably make room for their shared interest in naturalist studies. There are numerous science fiction books which have plots based on humans becoming smaller and having to deal with insects at their level.

Some examples are *The Insect Warriors* by Rex Dean Levie, *Atta* by Francis Rufus Bellamy, *Bug Park* by James P. Hogan, *The Micronauts* series by Gordon Williams, and *The Forgotten Planet* by Murray Leinster. *The Forgotten Planets* plot is twisted in that the insects are the size of men (or larger) on a planet "seeded" to prepare it for human habitation. Robert Asprin wrote *The Bug Wars*, a novel about war between reptiles and insects on an interplanetary scale. There are quite a few films about insects, or at least prominently featuring them. Widespread attitudes of revulsion and fear toward insects are often exploited by Horror and Science Fiction films through insect/ insect-like monsters (Them! is a famous early example), or by showing humans transformed into (The Fly) or attacked by insects. Another more positive type of insect film is animation with anthropomorphized insects as characters.

Identification of Insects

Most insects can easily be recognized to order, such as Hymenoptera (bees, wasps, and ants) or Coleoptera (beetles). However, insects other than Lepidoptera (butterflies and moths) are typically identifiable to genus or species only through the use of Identification keys and Monographs. Because the class Insecta contains a very large number of species (over 330,000 species of beetles alone) and the characteristics separating them are unfamiliar, and often subtle (or invisible without a microscope), this is often very difficult even for a specialist. Insect identification is an increasingly common hobby, with butterflies and dragonflies being the most popular.

Taxonomic Specialization

Many entomologists specialize in a single order or even a family of insects, and a number of these subspecialties are given their own informal names, typically (but not always) derived from the scientific name of the group:

- Apiology (or melittology) - bees
- Coleopterology - beetles
- Dipterology - flies
- Hemipterology - true bugs
- Lepidopterology - moths and butterflies
- Myrmecology - ants
- Orthopterology - grasshoppers, crickets, etc.
- Trichopterology - caddis flies

Organizations

Like other scientific specialties, entomologists have a number of local, national, and international organizations. There are also many organizations specializing in specific subareas.

- Amateur Entomologists' Society
- Deutsches Entomologisches Institut
- Entomological Society of America
- Entomological Society of Canada
- Entomological Society of Japan
- International Union for the Study of Social Insects
- Netherlands Entomological Society
- Royal Belgian Entomological Society
- Royal Entomological Society of London
- Société Entomologique de France

Museums

Here is a list of selected museums which contain very large insect collections.

Africa

- Natal Museum, Pietermaritzburg, South Africa

Europe

- World Museum Liverpool
- The Bugworld Experience, Liverpool
- Natural History Museum, London Natural History Museum
- Natural History Museum, Vienna Naturhistorisches Museum
- Natural History Museum, Paris Muséum national d'histoire naturelle
- Natural History Museum, Berlin Humboldt Museum
- Royal Museum for Central Africa, Brussels Royal Museum for Central Africa
- Natural History Museum, Leiden Natural History Museum, Leiden
- Natural History Museum, Sweden Swedish Museum of Natural History
- Natural History Museum, St. Petersburg Zoological Collection of the Russian Academy of Science

- Natural History Museum, Geneva
- The Bavarian State Collection of Zoology Zoologische Staatssammlung München
- Natural History Museum, Budapest Hungarian Natural History Museum

United States

- Academy of Natural Sciences, Philadelphia, Pennsylvania
- American Museum of Natural History, New York City, New York
- Auburn University Entomological Museum, Auburn University, Auburn, Alabama
- Audubon Insectarium, New Orleans, Louisiana
- Bohart Museum of Entomology, Davis, California
- California Academy of Sciences, San Francisco, California
- Carnegie Museum of Natural History, Pittsburgh, Pennsylvania
- Essig Museum, Berkely, California
- Field Museum of Natural History, Chicago, Illinois
- Florida Museum of Natural History, University of Florida, Gainesville, Florida
- Museum of Comparative Zoology, Cambridge, Massachusetts
- Natural History Museum of Los Angeles County, Los Angeles, California
- National Museum of Natural History, Washington, District of Columbia
- North Carolina State University Insect Museum, Raleigh, North Carolina
- Peabody Museum of Natural History, New Haven, Connecticut
- Texas A&M University, College Station, Texas
- University of Kansas Natural History Museum, Lawrence, Kansas

- University of Nebraska State Museum, Lincoln, Nebraska
- University of Missouri Enns Entomology Museum, University of Missouri, Columbia, Missouri

Canada

- Canadian Museum of Nature, Ottawa, Ontario
- Canadian National Collection of Insects, Arachnids and Nematodes, Ottawa, Ontario
- E.H. Strickland Entomological Museum, University of Alberta, Edmonton, Alberta
- Lyman Entomological Museum, Macdonald Campus of McGill University, Sainte-Anne-de-Bellevue, Quebec
- Montreal Insectarium, Montreal, Quebec
- Royal Alberta Museum, Edmonton, Alberta
- Royal Ontario Museum, Toronto, Ontario
- Newfoundland Insectarium, Reidville, Newfoundland and Labrador
- University of Guelph Insect Collection, University of Guelph, Guelph, Ontario

TIMELINE OF ENTOMOLOGY

Entomology, the scientific study of insects and closely-related terrestrial arthropods, has been impelled by the necessity of societies to protect themselves from insect-borne diseases, crop losses to insects, and insect-related discomfort, as well as by people's natural curiosity. This article traces the history of entomology.

History of Classification

Many different classifications were proposed by early entomologists. It is important to realise that whilst many early names survive they are at different levels in the phylogenetic hierarchy. For instance many families were first published as genera. An example is the genus *Mymar* proposed by Alexander Henry Haliday in 1829 is now the family Mymaridae.

19th Century

1800 – an arbitrary date but it was around this time that systematists began to specialise. There remained entomological polyhistors – those who continued to work on the insect fauna as a whole. From the beginning of the century, however, the specialist began to predominate, harbingered by Johann Wilhelm Meigen's *Nouvelle classification des mouches à deux aile* (New classification of the Diptera) commenced in the first year of the century. Lepidopterists were amongst the first to follow Meigen's lead. The specialists fell into three categories. First there were species describers, then specialists in species recognition and then specialists in gross taxonomy. There were however considerable degrees of overlap. Also then, as now, few could entirely resist the lure of groups other than their own, and this was especially true of those in small countries where they were the sole 'expert', and many famous specialists in one order also worked on others. Hence, for instance, many works which began as butterfly faunas were completed as general regional works, often collaboratively. "Man is born not to solve the problems of the universe, but to find out where the problem begins, and then to restrain himself within the limits of the comprehensible" Johann Wolfgang von Goethe *Conversations with Eckerman: Feb. 13, 1829*

1800

- Jean-Baptiste Pierre Antoine de Monet, Chevalier de Lamarck first expressed his views on evolution in lectures.
- The total number of species of insects described is estimated at not exceeding the figure of 20 000.

1801

- Publication of Jean Baptiste Pierre Antoine de Monet

de Lamarck. *Système des animaux sans vertèbres ou tableau général des classes, des ordres et des genres de ces animaux.* Paris:Deterville in English, 'System of invertebrate animals or general table of classes, orders and genera of these animals'

- Johan Christian Fabricius *Systema eleutheratorum* commenced. In a series of successive works to 1806 Johann Christian Fabricius developed a classificatory system based entirely on the structure of the mouthparts.
- Johann Karl Wilhelm Illiger's *Magazin für Insectenkunde*or Insect Magazine commenced in Brunswick (last issue 1856). One of the first insect magazines.
- Marie Françoise Xavier Bichat. *Traité d'anatomie descriptive*, in English, 'Treatise on descriptive anatomy published in Paris'. This work laid the foundations of a unified structural terminology.
- The Aurelian Society meets in London *"To form a complete and standard cabinet of the entomological productions of Great Britain:- to ascertain their names, uses, and distinctions: the places and times of their appearance, food economy and peculiarities :- and to point out to the public the readiest and most desirable methods of destroying such as possess properties that are inimical to the welfare of mankind"*
- First part of Pierre André Latreille's Historie naturelle, *générale et particulièredes crustacés et des insects* in English, 'General and particular natural history of crustaceans and insects', published in Paris by Dufart. 5817 pp. (14 volumes last published in 1805). In this and succeeding works (1806–1809, 1810, 1825, 1829 Latreille, following Linnaeus, divides the insects into Orders (at first these are called classes), invented the important family concept and also the tribus, proposed the families Asilidae, Muscidae, Syrphidae and

Tabanidae (Diptera) as supra generic and erected the insect orders: Parasita, Perliarae, Megaloptera and Panorpatae. This is an important work of higher classification.

- Lamarck and Gottfried Reinhold Treviranus, separately proposed term "biology" for the whole science of living things.
- Pietro Rossi becomes the world's first professor of entomology in Pisa.

1802

- Charles Nicolas Aubé born.

1803

- Johann Christian Fabricius *Rhyngotorum secundum ordines*...Reichard, Brunsvigae. Published.
- Adrian Hardy Haworth Lepidoptera Britannica (1803–1828) published. This was the most authoritative work on British butterflies and moths until Henry Tibbats Stainton 's *Manual* in 1857.
- Johann Rudolph Schellenberg *Genres des mouches Diptères représentés en XLII planches projettées, dessinées et expliquées par deux amateurs de l'entomologie* published in Zürich.
- Jakob Heinrich Laspeyres *Kritische Revision der neuen Ausgabe des systematischen Verzeichnisses von den Schmetterlingen der Wienergegend*. Braunschweig, Karl Reichard.

1804

- Jacob Hübner *Sammlung Europaischer Schmetterlinge*, in English, 'Collection of European Butterflies', published in Augsburg.
- Caspar Erasmus Duftschmid *Fauna Austriae*, in English, 'Austrian fauna' commenced publication.
- Jacob Sturm *Deutschlands Insecten, Käfer*, in English, 'German Insects, Beetles' commenced. This work was not completed until 1856. Like many

insect faunal works, this was planned to cover all orders but got no further than beetles.

- Thomas Marsham. *Entomologica Britannica sisterns Insects Britanniae Indigena Secundum Linneum disposita* in English, 'British Entomology comprising native British insects disposed (according to) Linnaeus' second edition. The first British beetle fauna this work treated Coleoptera only.
- Jean Frederick Hermann *Mémoire Aptérologique* published by F.L. Hammerin Strasbourg.

1805

- Ambroise Marie Francise Joseph Palisot de Beauvais 1805-1821 *Insectes Recueillis en Afrique et en Amerique*, in English, 'Insects found in Africa and America'
- Prodromus Entomology published

1806

- Carl August Dohrn born in Germany (States)
- Anders Gustaf Dahlbom (1806–1859) born in Sweden.
- Pierre André Latreille *Genera Crustaceorum et Insectorum Secundum Ordinem Naturalem in Familias Disposita, Iconibus, Exemplisque Plurimus Explicata* 4 vols. Paris 1257pp. commenced, completed 1809.

  Latreille wrote a series of works each revising his previous classification. All are important work of higher classification.
- Jacob Hübner. *Tentamen determinationis, digestionis atque demonstrationis singularum stirpium Lepidopterorum, peritis ad inspiciendum et disjudicandum communicatum,* published in Augsberg. In English, the title reads 'Preliminary examination. An attempt to fix, arrange and name the individual races of Lepidoptera to experts for examination and the expression of an opinion'. As

its title states, this was a discussion document. Inadvertently published, it led to subsequent nomenclatural confusion.

- Maximilian Spinola *Insectorum Liguriae* etc. published in Genoa, Italy.
- André Marie Constant Duméril, 1806 *Analitische Zoologie.* Published in both French and German. This was an important text for its methodology of classification. It is also an important work of higher classification.

1807

- Louis Jurine *Nouvelle Méthode de Classer les Hyménoptères et les Diptères*, in English, 'New method of classifying Hymenoptera and Diptera'. Published in Genève (Geneva).
- Birth year of three specialists in Diptera, Hermann Loew, Alexander Henry Haliday and Camillo Rondani
- Guillaume-Antoine Olivier *Le Voyage dans l'Empire Othoman, l'Égypte et la Perse.*
- 1807: Agostino Bassi (1773–1856) begins research on a disease of the silkworm caused by fungi, preceding work of Louis Pasteur.

1808

- Publication date of Johann Cristoph Friedrich Klug Die Blattwespen nach ihren Gattungen und Arten zusammengestellt. *Mag. Ges. Naturf. Freunde*, in English, 'The leaf wasps arranged according to their genera and species'.
- Guillaume-Antoine Olivier *Entomologie, ou Histoire naturelle des Insectes. Coléoptères.*

1809

- Lamarck's views on evolution were fully elaborated in his *Philosophie zoologique* in which he also arranged animals according to relationships and was first to employ the genealogical tree.

1810

- Guillaume-Antoine Olivier redefined the order Orthoptera (Latreille, 1796). This Order has been subject to constant revision.
- Franco Andrea Bonelli published *Observations Entomologique*.
- George Perry began *Arcana*, also titled *The museum of natural history*.

1811

- Gustaf von Paykull, *Monographia Histeroidum* Suecicae published in Upsalla. A monograph is a very complete work on a restricted subject, in this case Histeridae. This is one of the first entomological monographs.
- Kurimoto Masayoshi *Kurimoto's Iconographia Insectorum*or Insects of Japan published.

1812

- Birth of the entomologist Mary Ball.
- Volume 1 of the *Transactions of the Entomological Society of London* published (though the Society officially dates from 1832)

1813

- William Kirby erected the orders Trichoptera and Strepsiptera
- Jean Baptiste Boisduval, *Faune Entomologique de Madagascar* published.

1814

- William Elford Leach. *Zoological Miscellany* begun at London.
- Carl Fredrik Fallén, *Diptera Sueciae*, in English, Diptera of Sweden published at Lund. Fallén's 484 page "Dissertation" this specialist work was published between 1814 and 1825

1815

- William Elford Leach founded the orders Phasmida, Anoplura, Thysanura and Rhaphidides; the hemipterous families Pentatomidae, Coreidae, Belostomidae; the dipterous family Tipulidae and the hymenopterous family Chrysididae and published the first bibliography of entomology in *Brewster's Edinburgh Encyclopedia.*
- William Kirby and William Spence (entomologist), *Introduction to entomology or elements of the natural history of insects.* 4 vols. London, Longman 2430pp. This masterwork commenced in 1815 and was completed in 1826. It is an outstanding achievement: an entomology and a system of higher units in which Kirby was much influenced by MacLeay. By 1818 it was in its third edition, one of the most popular scientific works of all time.
- Johann Friedrich Eschscholtz embarks on the Russian expeditionary ship "Rurik" 1815-1822
- Publication of Tome 1 of Jean Baptiste Pierre Antoine de Monet de Lamarck's *Histoire naturelle des animaux sans vertèbres.,* completed with Tome 7 in 1822. Paris, Verdiere.

1816

- Jacob Hübner *Verzeichniss bekanuten schmetterlinge*, in English, Catalogue of known butterflies published at Augsberg.
- Marie Jules Cesar Lelorgne de Savigny, *Mémoires sur les animaux sans vertèbres* published at Paris.
- Henri Marie Ducrotay de Blainville *Prodrome d'une nouvelle distribution systématique du règne animal* Bull. Soc. Philom. Paris
- Royal Saxony School of Forestry established at Tharandt. The school taught entomology and later the dipterist Samuel Friedrich Stein became its Director.

- First issue of Oken 's *Isis, eine encyclopädische Zeitschrift, vorzüglich für Naturgeschichte, vergleichende Anatomie und Physiologie* published.

1817

- First volume of Thomas Say's *American entomology* published in Philadelphia. The pioneering work of American entomology. Some of the illustrations were undertaken by Titian Peale.
- Carl Rheinhold Sahlberg, Dissertatio entomologica insecta Fennica enumerans (Coleoptera) commenced. First work on the Coleoptera of Finland. Due to the efforts of Paykull, Leonard Gyllenhaal and Sahlberg the beetles of Scandinavia were better known than in any comparable area at this date.

1818

- Johann Wilhelm Meigen *Systematische Beschreibung der Bekannten Europäischen zweiflugen Insecten* 7 vols. Aachen and Hamm 2869pp., in English, Systematic Descriptions of known European Two-winged Insects commenced. The seven volumes spanned the years 1818-1830. *Nouvelle classification des mouches à deux ailes*. Paris. New classification of the Diptera had already appeared, published in Paris.
- Carl Gustav Carus *Lehrbuch der Zootomie*. Leipzig. Lehrbuchs are student texts. This one is on anatomy, mainly dissection.
- Friedrich Wilhelm Ludwig Suckow *Anatomisch-physiologische Untersuchungen der Insecten und Krustenthiere*. 70 S. mit 11 Kupfern. Engelmann, Heidelberg.

1819

- George Samouelle publishes *A nomenclature of British Entomology, or a catalogue of above 4000 species of the Classes Crustacea, Myriapoda, Spiders,*

Mites and insects intended as labels for cabinets of Insects, etc., alphabetically arranged at London.

1820

- Carl Fredrik Fallèn *Monographia Muscidum Suecia* commenced.
- Philogène Auguste Joseph Duponchel with Jean Baptiste Godart, in English, *Histoire Naturelle des Lépidoptères de France* commenced. This was the first really good French butterfly fauna.
- Johann Fischer von Waldheim *Entomographia Imperii Russici, Genera Insectorum Systematice Exposita et Analysi Inconographia Instructa* published in Moscow.
- Gustaf Johan Billberg *Enumeratio insectorum in museo Gust. Joh. Billberg. Typus Gadelianus.*

1821

- John Fleming *Insecta*. In: *Supplement to the fourth, fifth and sixth editions of the Encyclopædia Britannica, with preliminary dissertations on the history of the sciences*an important systematic work.

1822

- Jacob Johann Hagenbach *Insectorum Helvetiae exhibentia vel species novas vel nondum depictas.* Basle

1823

- Philipp Franz von Siebold begins natural history studies in Japan. Some of his many Japanese students take up entomology. The species they discovered were described in European journals without explicit attribution. They are honoured here. The invertebrate volume of Siebold's *Fauna Japonica* was undertaken by Wilhem de Haan
- Johann Wilhelm Dalman 1823, publishes *Analeceta Entomologica* Stockholm

1824

- John Curtis with Alexander Henry Haliday and Francis Walker *British Entomology, being illustrations and descriptions of the genera of insects found in Great Britain and Ireland; containing coloured figures from nature of the most rare and beautiful species, and in many instances of the plants upon which they are found* 16 volumes 193 Folios 770 coloured plates commenced. A masterpiece of the engraver's and colourist's art, this work includes new classifications and many new species descriptions. The work falls between the illustrated works with little text of the previous years and the minimally illustrated revisions and monographs of later years.
- Leon Jean Marie Dufour, *Recherches anatomiques sur les Carabiques et sur plusiers autres Coléoptères,* published at Paris.

1825

- Pierre André Latreille *Familles Naturelles du Règne Animal* published. In this publication the name Insecta is finally restricted to the hexapod arthropods here called Class Hexapoda.
- Pierre François Marie Auguste Dejean *Species Général des Coléoptères* commenced (five volumes devoted to Carabidae) C. Aubé wrote the sixth on Hydrocanthares completing the work in 1838. It is a masterpiece of descriptive entomology.

1826

- Carl Fredrik Fallén *Hemiptera Sueciae.*
- William Kirby erected the order Aphaniptera for the fleas.
- William Sharp MacLeay *Horae entomologicae*

1827

- Pierre Léonard Vander Linden commenced

Observations sur les Hyménoptères d'Europe de la famille des Fouisseurs.

1828

- James Francis Stephens *Illustrations of British Entomology* published.
- Christian Gottfried Daniel Nees von Esenbeck *Monographie der Ichneumone.* A seminal work on Ichneumonidae in two volumes.
- Karl Ernst von Baer *Uber Entwickelungsgeschichte der Thiere* published. Baer was an embryologist whose "biogenetic law" of recapitulation was a major reference point for 19th century evolutionary theory.
- Pierre Boitard publishes *Manuel d'entomologie, ou Histoire naturelle des insectes : contenant la synonymie et la description de la plus grande partie des espèces d'Europe et des espèces exotiques les plus remarquables* (Roret, Paris)

1829

- Jean Baptiste Boisduval and John Eatton Le Conte published first installment of *Histoire général et iconographie des lepidoptérès et des chenilles de l'Amerique septentrionale* at Paris. (English title: *General history and illustations of the Lepidoptera and moths (sic) of North(ern) America.*) Publication of installments continued through 1837.
- Georges Cuvier, *Regne Animalium*, in English, *The Animal Kingdom*, published by Chez Deterville at Paris.
- Hermann Burmeister *De Insectorum Systemate Naturali* (Diss., Grunert, Halle)
- Johann Friedrich von Eschscholtz *Zoölogische atlas enthaltend abbildungen neuer thierasten während Kotzebues Zweiter reise um die welt gesammelt*, in English, Zoological Atlas containing illustrations of the animals collected during Kotzebues etc..

Second travels around the world 1823-1826 published at Berlin.

- Pierre Lyonet *Anatomie de différentes espèces d'insectes* published Paris
- Pierre François Marie Auguste Dejean, Jean Baptiste Boisduval and Charles Nicholas Aubé. Iconographie et histoire naturelle des Coléoptères d'Europe.
- Alfred Brehm 's *Tierleben* (English title: Brehm's Life of Animals) commenced publication. Very popular it was published in many expanded editions the six volumes published between 1864 and 1869 was titled *Illustrirtes Thierleben*, Ernst Ludwig Taschenberg wrote the entomological sections.
- Christian Friedrich Freyer published Beitrage zur Geschichte europaischer Schmetterlinge mit Abbildungen nach der Natur.

1830

- Controversy between Georges Cuvier and Étienne Geoffroy Saint-Hilaire on the fixity of types.
- Maximilian Perty *Delectus Animalium Articulatorum quae in itinere per Brasiliam Annis MDCCCXVII - MDCCCXX Iussu et Auspiciis Maximiliani Josephi I. Bavariae Regis Augustissimi, percato collegerunt Dr J.B. de Spix et Dr. C.F. Ph. de Martius.* (completed 1834)

1831

- Félix Édouard Guérin-Méneville. *Magasin de Zoologie, d'Anatomie Comparée et Palaeoentomologie*, in English, Magazine of Zoology, Comparative Anatomy and Palaeoentomology commenced publication.
- George Newport, Nervous system of *Sphinx ligustri*. London
- Jean Guillaume Audinet-Serville published *Revue méthodique de l'ordre des Orthoptères*

- Thaddeus William Harris prepared a catalogue of insects for Edward Hitchcock's *Report on the Geology, Mineralogy, Botany and Zoology of Massachusetts.*
- Jean-Baptiste Deyrolle, natural history dealer opened for business at 23, Rue de la Monnaie, Paris later assisted by his son Achille Deyrolle. Deyrolle et fils published many entomological works.

1832

- Société Entomologique de France founded at Paris. The first president was Amédée Louis Michel Lepeletier de Saint Fargeau
- Hermann Burmeister, *Handbuch der Entomologie*, in English, *Handbook of entomology*, commenced publication. An important work of higher classification.
- Jules Dumont d'Urville Voyage de l'Astrolabe. *Faune entomologique de l'Océanie par le* Dr Boisduval. *Tome 1 : Lepidoptéres* (1832) published; Tome 2 Coléoptères, Hémiptères, Orthoptères, Névroptères, Hyménoptères et Diptères was published in 1835.
- Felipe Poey *Centurie de Lépidoptères de l'Ile de Cuba* published at Paris.
- Hermann Friedrich Stannius *Beiträge zur Entomologie, besondere in Bezug auf Schlesien, gemeinschaftlich mit Schummel*. Breslau, published.

1833

- Alexander Henry Haliday, *An essay on the classification of the parasitic Hymenoptera of Britain which correspond with the Ichneumones minuti of Linnaeus.* Ent. Mag. First part of a seminal work on Hymenoptera erecting major taxa (phylogenetic divisions).
- Carl Johan Schönherr, Leonard Gyllenhaal and Carl Henrik Boheman. *Genera and species of Curculionidae.* 8 vols. Paris, 1833-1844 commenced.
- Francis Walker, *Monographia Chalcidum*. London,

1833–1842, commenced. Much of this work was collaborative with Alexander Henry Haliday who was the sole author of the sectional diagnoses.

- Louis Alexander Auguste Chevrolat, Coléoptères de Mexique.
- Pierre François Marie Auguste Dejean *Catalogue des Coléoptères*. Paris.
- George Robert Gray *The Entomology of Australia: Monograph of the Genus Phasma* Volume 1 Masterly work on Phasmidae. The first work on this group since Caspar Stoll (1780). Very often a monograph is separated from its successor by thirty to fifty years.
- Thaddeus William Harris *Report on the Geology, Mineralogy, Botany, and Zoolopgy of Massachusetts* published.
- Julius Theodor Christian Ratzeburg *Die Forstinsekten*, Berlin 1837–1844 commenced. This is a founding work of forest entomology.

1834

- Entomological Society of London founded.
- François Louis de la Porte, comte de Castelnau *Études entomologique. Introduction à l'entomologie* in

Suites à Buffon. Paris 1834-1863.

- Justin Pierre Marie Macquart *Histoire Naturelle des Insectes Diptères* published.

1835

- Constantin Wesmael commences publication of *Monographie des Braconides de Belgique*

1836

- Wilhelm Ferdinand Erichson published *Die Kafer der Mark Brandenburg*.
- Charles Athanase Walckenaer and Paul Gervais commenced *Histoire naturelle des insectes*.

1837

- Oswald Heer published *Die Kafer der Schweiz*
- Second edition of John Curtis "*A guide to the arrangement of British insects being a catalogue of all the named species hitherto discovered in* Great Britain *and* Ireland " published. The list contains 1500 generic and 15,000 specific names.
- Carl Gustaf Mannerheim (naturalist) *Mémoire sur quelques genres et espèces de Carabiques* published.
- Achille Rémy Percheron publishes *Bibliographie entomologique* listing more than 5 000 authors and 500 anonymous contributions.

1838

- Johan Wilhelm Zetterstedt *Insecta Lapponica Descripta*. Leipzig: Voss 1145pp (1838–1842) commences publication.
- Benjamin Dann Walsh left England for the United States of America. He was a pioneer of applied entomology.

1839

- John Obadiah Westwood. First part of *An introduction to the modern classification of insects*. (1839–1840) published.
- John Forbes Royle *Illustrations of the Botany and Other Branches of the Natural History of the Himalayan Mountains, and of the Flora of Cashmere* published. This work resembles 18th century works in its sumptuous illustration.
- Meyers Konversations-Lexikon commenced. Entomology was a favourite topic.

1839

- Etienne Mulsant. *Histoire naturelle des Coléoptères de France*. 13 pts. Paris and Lyon, 1839-1863 published.
- James Francis Stephens *Manual of British*

Coleoptera or Beetles This work contains diagnoses of 3462 species and remained the standard work until the appearance of the *Handbook of the Coleoptera* by H.E. Cox was published in 1875. An instance of the logevity of some taxonomic works.

- Alexander Henry Haliday, *Hymenopterorum Synopsis and Methodum Fallenii ut plurimum accomodata*. This work on classification was privately printed. Very few copies survive. The same is true of many important works of other authors.
- Edward Newman founded *The Entomologist* (an insect magazine) at London.
- Giuseppe Gené De quibusdam insectis Sardiniae novis aut minus cognitis. *Fascicle 2. Mem. R. Accad. Sci. Torino* (2)1: 43-84.

1840

- Jean Baptiste Alphonse Dechauffour de Boisduval *Genera et index methodicus Europaeorum lepidopterorum*, published in Paris.
- Theodor Becker born.
- Jan Dzierson designed a movable-frame beehive, the first of many apicultural advances he initiated.

1841

- William Wilson Saunders becomes president of the Entomological Society of London.

1842

- Jean Victoire Audouin publishes *Histoire des insectes nuisibles à la vigne et particulièrement de la Pyrale*.
- Jules Pierre Rambur publishes *Histoire naturelle des insectes*.

1843

- Johann Heinrich Kaltenbach. Published *Monographie der familien der Pflanzenhause*at Aachen. A monograph on Thysanoptera.

- Henry Noel Humphreys, an illustrator and John Obadiah Westwood *British Moths and their Transformations* London: William Smith, 1843-1845. 2 Volumes. Humphreys orchestrated the overall work and provided all illustrations. Westwood provided the various scientific descriptions.
- Anders Gustaf Dahlbom *Hymenoptera Europaea Praecipue Borealia* 1-2. Lundberg, Lund 1008 pp. commenced publication (completed 1854)
- Alcide d'Orbigny began *Dictionnaire Universel d'Histoire Naturelle*. This work was finished in 1849.

1844

- Thomas Workman born.

1845

- Netherlands Entomological Society founded.

1846

- Smithsonian Institution founded in Washington, D.C.

1847

- Edmond Ruffin publishes a study of the life history and cultural control of the "Corn or fly-weevil" (*Sitotroga cerealella* Olivier) in the *Farmers Register*. Fourteen years later Edmond Ruffin pulled the lanyard on one of the columbiads off Morris Island the first shot fired in the American Civil War but not the first entomologist to engage in violent politics.
- Hercule Nicolet *Essai sur une classification des insectes aptères, de l'ordre des Thysanoures*.

1848

- Carl Theodore Ernst von Siebold establishes the phylum Arthropoda.
- Maximilien de Chaudoi r.*Memoire sur la famille des Carabiques*. 6 parts, published at Moscow.
- John Obadiah Westwood *The Cabinet of Oriental Entomology*

- Charles Jean Baptiste Amyot 1848. *Entomologie Française. Rhynchotes. Méthode mononymique* Paris, J.-B. Bailliere, Libraire de l'Acad. An important step forward in the classification of Coleoptera and an important work of higher classification.

1849

- Étienne Mulsant *Species des Coléoptères Trimères sécuripalpes.* 2 parts. Lyon, 1849-1853.
- Hope Chair of Zoology established at Oxford by Frederick William Hope and occupied by John Obadiah Westwood.
- Alexander von Humboldt. *Ansichten der Natur.* Stuttgart.

1850

- Edmond de Sélys Longchamps Revue des odonates ou Libellules d'Europe. *Mémoires de la Société Royale des Sciences de Liége* 6:1–408.
- Victor Ivanovitsch Motschulsky *Die Kaefer Russlands.* I. *Insecta Carabica.* Russian beetles, Carabidae, Moscow: Gautier, published.

1851

- Johann Fischer von Waldheim and Eduard Friedrich Eversmann publish *Lépidoptères de la Russie* (vol. 5 of Johann Fischer von Waldheim. *Entomographia Imperii Rossici.* Seminal work on Russian Lepidoptera.
- Louis Agassiz. *On the classification of insects from embryological data.* Washington, published.
- Francis Walker. *Insecta Britannica Diptera* 3 vols. London 1851-6. The characters and synoptical tables of the order by Alexander Henry Haliday made this a seminal work of Dipterology.
- Hans Hermann Behr emigrates from Germany to California.

1852

- Achille Guenée *Histoire naturelle des insectes. Species general des Lépidoptères*. Paris, 1852–1857, published.

1853

- Leopold Heinrich Fischer publishes *Orthoptera Europaea* and pernounces himself gay with Samuel de Champlain. Lipsiae, (Leipzig) G. Engelmann, 1853. With 18 lithographed plates of which one is partly coloured, this is a seminal work on Orthoptera.
- Frederick Smith *Catalogue of Hymenopterous Insects* (7 parts, 1853–1859)

1854

- Jean Théodore Lacordaire, *Histoire Naturelle des insectes. Genera des Coléoptères*. 9 vols published at Paris, 1854–1869 (completed by Félicien Chapuis, vols. 10–12, 1872–1876).
- Carl Ludwig Koch *Die Pflanzenlause*, etc. Nurnburg commenced – completed 1857.
- Ignaz Rudolph Schiner *Diptera Austriaca. Aufzahlung aller im Kaiserthum Oesterrich bisher aufgefundenen Zweifluger*, 1–4 *Verh. Zool. Bot. Ver. Wien*. 4–8 263 pp. (1854–1858) commenced.
- Émile Blanchard (1819–1900) writes *Zoologie agricole où il présente les espèces nuisibles*, a work on pest species. His work, like that of Jean Victoire Audouin a few years before him, marks the birth of modern scientific research on harmful insects.
- Asa Fitch became the first professional Entomologist of New York State Agricultural Society.

1855

- Camillo Rondani *Dipterologiae Italicae Prodromus* 1–5. Parma: Stochi 1146 pp. commenced (completed 1862)

- Eduard Friedrich Eversmann *Les Noctuelites de la Russie* first volume (completed 1859)
- Henry Tibbats Stainton, Philipp Christoph Zeller, John William Douglas and Heinrich Frey *The Natural History of the Tineina* 13 volumes, 2000 pages. One of the most significant lepidopterological works of the century-The Natural History of the Tineinae, is a monumental 13 monographic work.

1856

- Baron Carl Robert Osten Sacken became Russian General Counsel in New York.
- Ernest Candèze *Monographie of Elateridae* (four volumes, Liege, 1857–1863) commenced.

1857

- William Chapman Hewitson 1857–76 *Illustrations of New Species of Exotic Butterflies*. London, 1857–1861 commenced.
- Pierre Nicolas Camille Jacquelin Du Val (1828–1862) *Manuel entomologique. Genera des Coléoptères d'Europe* commenced (finished 1868 by Léon Fairmaire).

1858

- Henri Louis Frederic de Saussure *Mélanges Hyménoptérologiques*. 1.67 p., 1 pl – also included in *Mémoires de la Société de Genève*.
- Ludwig Redtenbacher publishes *Fauna Austriaca. Die Käfer, nach der analytischen Methode bearbeitet*

1859

- Herman von Heinemann *Die Schmetterlinge Deutschlands und der Schweiz* in English, "Butterflies of Germany and Switzerland". Completed 1877 The second volume on microlepidoptera was especially important.
- Charles Darwin *Origin of Species* London. Entomologists had differing views of this work.

Best known for his theory of evolution through natural selection Darwin was also a keen entomologist.

- Leander Czerny born.
- The Museum of Comparative Zoology at Harvard founded by Louis Agassiz.
- The Entomological Society of Philadelphia was established. In 1867, it was renamed the American Entomological Society, the earliest national organization in the biological sciences in the United States of America.
- Pierre Millière commenced *Iconographie et description de Chenilles et Lépidoptères inédit*not completed until 1874.

  His collections of Macrolepidoptera and Pyralidae are in Palais Coburg in Vienna, there are some of his Microlepidoptera in the Natural History Museum, Leiden but the bulk of his Microlepdoptera collection is in the Muséum national d'Histoire naturelle in Paris.
- The Russian Entomological Society is founded with Academician K.M. Baer as the first president.

1860

- John Curtis *Farm Insects being the natural history and economy of the insects injurious to the field crops of Great Britain and Ireland with suggestions for their destruction Glasgow*, Blackie. Seminal work on economic entomology.
- Giovanni Passerini 's *Gli afidi con un prospetto dei generi ed alcune specie nuove Italiane* published at Parma
- Theodor Becker born.
- Franz Xaver Fieber *Die europäischen Hemiptera*

1861

- John Lawrence LeConte *Classification of the*

Coleoptera of North America published LeConte was the most important American entomologist of the century.

- Otto Staudinger and Maximilian Ferdinand Wocke *Catalog der Lepidopteren Europas*
- Carl Gustav Carus *Natur und Idee oder das Werdende und sein Gesetz. Eine philosophische Grundlage für die specielle Naturwissenschaft.* Wien: Braunmüller. Important science philosophical work.
- Jørgen Matthias Christian Schiødte. *De Metamorphosi Eleutheratorum Observationes. Bidrag til Insekterns Udviklingshistorie. Naturhistorisk Tiddsskrift* commenced. 13 parts completed 1883. Seminal work on larvae of Coleoptera.
- Museum Godeffroy opened in Hamburg.

1863

- Karl Friedrich Wilhelm Berge *Schmetterlingsbuch* Third edition of this 194 page popular classic work with 49 hand-coloured plates. Berge also wrote *Käferbuch* (1841). This work became an enduring classic. William Forsell Kirby used its 61 plates for his 1889 European Butterflies and moths. It was as *B. J. Rebel's edition of Berge's Schmetterlingsbuch* still a bestseller in 1949 and is now available as a CD.
- Entomological Society of Canada founded.
- Sylvain Auguste de Marseul *Catalogue des coléoptères d'E[illegible] du bassin de la Mediterranée en Afrique et en [illegible]*. Deyrolle, Paris., published.

1864

- Francis Polkinghorne Pascoe Longicornia Malayana; or a descriptive catalogue of the species of the three longicorn families Lamiidae, Cerambycidae and Prionidae collected by Mr. A. R. Wallace in the

Malay Archipelago. *Trans. Entomol. Soc. London* commenced. This work was finished in 1869

- Alexander Walter Scott, 1864-6 Australian Lepidoptera with their Transformations. A beautifully illustrated (by Harriet and Helena Scott) seminal work of Australian Entomology.

  First appearance of the *Entomologist's Monthly Magazine* in England together with the reappearance of the *Entomologist* indicates a surge of entomology in England.
- *Zoological Record* started in London. Continues work of Hagen, but includes taxa other than insects.
- Carl Stål Hemiptera Africana. 1–4, Holmiae, Stockholm. [in Latin, textual descriptions, keys to genera] 1864–1866.
- Frédéric Jules Sichel wrote *Catalogus specierum generis Scolia* ... an important text on Hymenoptera.

1865

- Alfred Russel Wallace On the Phenomena of Variation and Geographical Distribution as Illustrated by the Papilionidae of the Malayan Region (vol. 25 *Trans. Linn. Soc. Lond*). Seminal biogeographic and evolutionary work essentially laying out the principles of allopatric speciation.
- Cajetan Freiherr von Felder, Rudolf Felder & Alois Friedrich Rogenhofer *Fregatte Novara. Lepidoptera.* 1–3. commenced.
- Robert Mac Lachlan Trichoptera Britannica; a monograph of the British species of Caddis-flies. *Trans. Entomol. Soc. Lond.* (3) 5: 1–184
- Samuel Hubbard Scudder An inquiry into the zoological relation of the first discovered traces of fossil neuropterous insects in North America; with remarks on the different structure of wings of living Neuroptera. – *Mem. Boston soc. nat. hist.*, 1: 173–192 1865-1867

1866

- Josef Mik Beitrag zur Dipterenfauna des österreichischen Küstenlandes. *Abh. Zool.-Bot. Ges. Wien* 16:301–310, published. Mik's first work on the Diptera.
- Carl Eduard Adolph Gerstaecker commenced *Arthropoda*, in *Klassen und Ordnungen des Thierreichs*, (Section Arthropoda, in classes and orders of the Animals) 1866. The work was finish in 1893.

1867

- Johannes Winnertz *Beitrag zu einer Monographie der Sciarinen*. Vienna.

1868

- August Emil Holmgren *Hymenoptera, species novas descripsit. Kongliga Svenska Fregatten Eugenies resa omkring Jordeni under befäl af C.A. Virgen Aren 1851–53*. II Zoologi 1 Insecta pp. 391–442 pl. viii.
- Charles Valentine Riley appointed to the office of entomologist of the State of Missouri.
- Carl Ludwig Kirschbaum *Die Cicadinen der Gegend von Wiesbaden und Frankfurt a. M. nebst einer Anzahl neuer oder schwer zu unterscheidender Arten aus anderen Gegenden Europas*

1869

- 1869: Tord Tamerlan Teodor Thorell publishes *European Spiders* followed, by (in 1870) *Synonymy of European Spiders*, two outstanding works which are significantly progress the taxonomy of the spiders.
- La Società Entomologica Italiana founded.
- Louis Pandelle Études monographique sur les staphylins européens de la tribu des Tachyporini

Erichson. *Annales de la Société Entomologique de France*.

- Ferdinand Ferdinandovitsch Morawitz Die Bienen des Gouvernements von St. Petersburg *Trudy Russkago éntomologicheskago obshchestva* 6: 27–71.

1870

- Thomas Ansell Marshall *Ichneumonidium Brittanicorum Catalogus*. London

1871

- Enrico Verson (1845–1927) founded the world's first silkworm experimental station in Italy.

1872

- Ferdinand Heinrich Hermann Strecker *Lepidoptera, Rhopaloceres and Heteroceras, Indigenous and Exotic; with Descriptions and Colored Illustrations* published.

1874

- Robert Mac Lachlan publishes the first volume of *Monographic revision and synopsis of the Trichoptera of the European fauna* (two volumes, 1874 et 1880).
- Achille Costa 1874. *Fauna Salentina*. Tip. Ed. Salentina, Lecce, Italia.

1875

- Antonio Curò *Saggio di un Catalogo dei Lepidotteri d'Italia* commenced (completed 1889).
- Eugène Anselme Sébastien Léon Desmarest *Hymenoptéres* in *Encyclopédie d'Histoire Naturelle*.

1876

- Augustus Radcliffe Grote published *The effect of the glacial epoch upon the distribution of insects in North America*.

- Cyrus Thomas appointed state entomologist of Illinois.
- Félicien Henry Caignart de Saulcy Species des Paussides, Clavigérides, Psélaphides & Scydménides de l'Europe et des pays circonvoisins. *Bull. Soc. Hist. Nat. Metz* 14: 25–100, published. Saulcy specialised in cave fauna.
- Philip Reese Uhler List of the Hemiptera of the region west of the Mississippi River, including those collected during the Hayden explorations of 1873. *Bulletin of the United States Geological and Geographical Survey of the Territories* 1:267–361 published.

1878

- Charles Valentine Riley (1843–1895) organized the first government agricultural entomology service in the United States of America.

1879

- *Biologia Centrali-Americana* (1879–1915) commenced by Osbert Salvin and Frederick DuCane Godman This was a 52 volume encyclopedia on the natural history of Central America.
- *Souvenirs entomologiques* by Jean-Henri Fabre commenced (ended 1907).
- Adolfo Targioni Tozzetti. *Notizie e indicazioni sulla malattia del pidocchio della vite o della fillossera (Phylloxera vastatrix).* Roma. Tipografia Eredi Botta

1880

- Edouard Piaget publishes *Les pédiculines: Essay monographique*. Brill, Leiden

1881

- Matthew Cooke *Treatise on the Insects Injurious to Fruit and Fruit Trees of the State of California, and Remedies Recommended for Their Extermination*. Sacramento: State Office: J. D.

Young, Supt. State Printing, 1881. A pioneering work of American entomology.

- John Henry Comstock became professor of entomology at Cornell University.
- Henri Gadeau de Kerville *Les Insectes phosphorescents : notes complémentaires et bibliographie générale (anatomie physiologie et biologie) : avec quatre planches chromolithographiées*, Rouen, L. Deshays, published.

1882

- Peter Cameron *A Monograph of the British Phytophagous Hymenoptera 1893* Ray Society commenced. A four volume work completed in 1893.

1883

- Friedrich Moritz Brauer *Die Zweiflügler des Kaiserlichen Museums zu Wien*. III. Systematische Studien auf Grundlage der Dipteren-Larven. Seminal work of Dipterology.
- Charles Lionel Augustus de Nicéville with George Frederick Leycester Marshall began Butterflies of India, Burmah and Ceylon, a three volume work completed in 1890.

1884

- Nicholas Mikhailovich *Les Lépidoptères de la Transcaucasie. Ire Partie*. In: *Mémoires sur les Lépidoptères*. Stassulewitsch, St.-Pétersbourg, Vol. 1 (1–92, pl. 1-50.
- Joannes Charles Melchior Chatin *Morphologie comparée des pièces maxillaires, mandibulaires et labiales chez les insectes broyeurs*

1886

- Deutsches Entomologisches Institut (DEI, German Entomological Institute) founded.
- Carl Gustav Alexander Brischke published *Die Hymenopteren des Bernsteins.*

- Mariano de la Paz Graëlls y de la Aguera *Entomología Judicial. Rev. Progr. Cienc. Exac. Fís. Nat. Madrid, 21:458–471.* A founding work of forensic entomology.

1898

- Liverpool School of Tropical Medicine founded, the first of its kind in the world. Between 1898 and 1913 the School despatched 32 expeditions to the tropics, including Sierra Leone, the Congo and the Amazon.
- Fritz Müller publishes Über die Vortheile der Mimicry bei Schmetterlingen in *Zoologischer Anzeiger* 1, 54–55.

1889

- The Entomological Society of America (ESA) founded.
- William Lucas Distant *A monograph of oriental* Cicadidae commenced, a seven part monograph with fifteen handcoloured lithographed plates finished in 1892.
- Julius Weise *Naturgeschichte der Insecten Deutschland. Erste Abtheilung Coleoptera. Sechster Band. Berlin, Nicolaische Verlags-Buchhandlung R. Stricker* published.

1890

- Edwin Felix Thomas Atkinson published *Catalogue of the Insecta*. Order Rhynchota (pars).

1891

- G. E. Grumm-Grshimailo Le Pamir et la faune lépidoptérologique. Mém. lépidop. Ed. N.M. Romanoff published
- Louis Gabriel d'Antessanty published *Catalogue des Hémiptères-Hétéroptères de l'Aube* Dufour-Bouquot Plaquette Grand In-8 Broché Troyes.
- Sigmund Exner *Die Physiologie der facettierten*

Augen von Krebsen und Insekten describes the compound eye physiology of insects and crustaceans.

1892

- Margaret Fountaine begins world trips collecting butterflies.
- *Miscellanea Entomologica* established.

1893

- George Francis Hampson published *The Lepidoptera Heterocera of Ceylon* (1893) as parts 8 and 9 of *Illustrations of Typical Specimens of Lepidoptera Heterocera of the British Museum*. He then commenced work on *The Fauna of British India, Including Ceylon and Burma*. Moths *(4 vols 1892–1896).*
- William Harris Ashmead publishes Monograph of the North American Proctotrupidae in the *Bulletin of the U.S. National Museum*
- Philip Powell Calvert publishes *Catalogue of the Odonata (dragonflies) of the Vicinity of Philadelphia, with an Introduction to the Study of this Group* a model for later regional studies.
- Eleanor Anne Ormerod *Manual of injurious insects with methods of prevention and remedy for their attacks to food crops, forest trees and fruit: to which is appended a short introduction to entomology* published.
- Maurice Noualhier 1893. Voyage de M. Ch. Alluaud aux iles Canaries (Novembre 1889 – Juin 1890). 2e Memoire. Hémiptères Gymnocerates & Hydrocorises. *Annales de la Société Entomologique de France* 52:5–18.

1894

- Jean Pierre Mégnin publishes *La Faune des Cadavres. Applications de l'Entomologie à la Médecine Légale*. Corpse Fauna: Application of

Entomology to Legal Medicine. Ency-clopedie Scientifique des Aides-Memoires, Masson et Gauthier-Villars, Paris, 214 pp.

- Ferdinand Kowarz 1894 *Catalogus insectorum faunae bohemicae*. -II. *Fliegen* (Diptera) published.
- Bureau of Entomology (U.S.A.) founded.
- Ernst Haeckel *Die systematische Phylogenie*- "Systematic Phylogeny"), published.

1895

- Alpheus Spring Packard First volume of Monograph of the Bombycine Moths of North America was published, (parrs 2 and 3 1905 and 1915). Packard was a vocal proponent of the Neo-Lamarckian theory of evolution.

1897

- Charles E. Woodworth born.
- Charles Thomas Bingham *The Fauna of British India, Including Ceylon and Burma. Hymenoptera.* 1. *Wasps and bees* published.
- Royal Museum for Central Africa established.

1898

- Pater Gabriel Strobl Fauna diptera Bosne, Hercegovine I Dalmacie. *Glasn. Zemalj. Muz. Bosni Herceg.* 10: 87–466, 562–616. In Serbian this was the first in a series of works on the Diptera af the Balkans.
- The French scientist Paul-Louis Simond established the rat flea as the vector of bubonic plague.

1900

- Walter Reed, a United States Army major, was appointed president of a board "to study infectious diseases in Cuba paying particular attention to yellow fever." He concurred with Carlos Finally in identifying mosquitoes as the agent.

- Ignacio Bolívar y Urrutia publishes *y Catálogo sinóptico de los ortópteros de la fauna ibérica.*
- Kálmán Kertész, Mario Bezzi, Paul Stein (entomologist) and Theodor Becker published the first part of a Palaearctic Catalogue of Diptera *Katalog der Paläarktischen dipteren* in Budapest.

1901

- William Francis de Vismes Kane *A catalogue of the Lepidoptera of Ireland*-the third (and first comprehensive) catalogue of the Irish macrolepidoptera.
- Augustus Daniel Imms *General textbook of Entomology* published. 10th revised edition (1977) still one of the most widely used of all insect texts.
- Thomas Hunt Morgan is the first to conduct genetic research with the fruit fly *Drosophila melanogaster*. In the Fly Room at Columbia University.

1902

- Ronald Ross gained Nobel Prize for Medicine for his discovery that malaria is carried by mosquitoes. The awarding committee made special mention of the work of Giovanni Battista Grassi on the life history of the Plasmodium parasite.
- Charles W. Woodworth *A List of the Insects of California* published.
- Philogene Auguste Galilee Wytsman started *Genera Insectorum*, a multi-authored series that consisted of 219 issues, the last occurring in 1970.
- Otto Schmiedeknecht *Opuscula Ichneumonologica*. Blankenburg.
- William Morton Wheeler appointed curator of invertebrate zoology in the American Museum of Natural History, New York

1905

- Adolfo Lutz Beitraege zur Kenntniss der

brasilianischen Tabaniden. *Rev. Soc. Sci. São Paulo* 1: 19-32, published

- Raphaël Blanchard *Les moustiques. Histoire naturelle et médicale* Paris, F.R. de Rudeval, published.

1906

- Adalbert Seitz, *Gross-Schmetterlinge der Erde* commenced. This vast work on Lepidoptera was published in German, English and French. It contained colour plates of all important species.

1907

- William Lundbeck *Diptera Danica. Genera and species of flies Hitherto found in Denmark* commenced.
- Herbert Druce On Neotropical Lycaenidae, with Descriptions of New Species. Proceedings of the Zoological Society of London.
- He[illegible]y Christopher McCook *Nature's Craftsmen: Popular Studies of Ants and Other Insects.*
- Lajos Abafi *Magyarország lepkéi* (butterflies of Hungary) 1907.

1908

- Edmund Reitter *Fauna Germanica - Die Käfer des Deutschen Reiches* commenced. This five volume masterwork remains in use today, almost 100 years from its inception.

1909

- George Henry Verrall *Stratiomyidae and succeeding families of the Diptera Brachycera of Great Britain - British flies* published.
- Carlos Chagas observed the peculiar infestation of rural houses in Brazil with Triatoma, a "kissing" bug, later demonstrating that it was the vector of *Trypanosoma cruzi*, and he was able to prove experimentally that it could be transmitted to

marmoset monkeys that were bitten by the infected bug. His description of the new disease was to become a classic in medicine and brought him domestic and international distinction.

- Charles Nicolle reasoned that it was most likely lice that were the vector for epidemic typhus. He tested his theory by infecting a chimpanzee with typhus, retrieving the lice from it, and placing it on a healthy chimpanzee. Within 10 days the second chimpanzee had typhus as well.
- Antonio Berlese Volume I of *Gli insetti loro organizzazione, sviluppo, abitudini e rapporti con l'uomo* (Volume 2 1925).
- Arnold Pagenstecher *Die geographische Verbreitung der Schmetterlinge* published in Jena.
- Foundation of *Journal of Entomology* by Charles Fuller Baker, one of the first revues of economic entomology.

1910

- Gilbert John Arrow published the first volume of *The Fauna of British India, Including Ceylon and Burma*. Lamellicornia 1. Cetoniinae and Dynastinae. Arrow wrote five volumes of this classic work.
- Hans Ferdinand Emil Julius Stichel Lepidoptera Rhopalocera. Fam. Riodinidae. Published in J. Wytsman *Genera Insectorum* 112A completed 1911).
- Hans Fruhstorfer published Family Pieridae in Adalbert Seitz' *Macrolepidoptera of the world*

1912

- Per Olof Christopher Aurivillius wrote Part 39 of *Catalogus Coleopterorum* Cerambycidae: Cerambycinae (1912). Aurivillius worked on world insects.

1913

- Charles Paul Alexander *A synopsis of part of the*

Neotropical Crane-flies of the subfamily Limnobinae (Tipulidae).

- Erwin Lindner joined the State Museum of Natural History Stuttgart.
- Otto Kröber Therevidae. Genera. Ins published.
- Karl Eckstein *Die Schmetterlinge Deutschlands mit besonderer Berücksichtigung der Biologie* commenced (finished 1933).
- G.D. Hale Carpenter joined the London School of Hygiene and Tropical Medicine, and took the Doctor of Medicine examination in 1913 with a dissertation on the tsetse fly (Glossina palpalis) and sleeping sickness.

1914

- Friedrich Georg Hendel Die Arten der Platystominen. *Abh. Zool.-Bot. Ges. Wien* 8 (1): 1-409, 4 pls. published
- Filippo Silvestri Contribuzione alla conoscenza dei Termitidi e Termitofili dell'Africa occidentale. *Bollettino del Laboratorio di Zoologia General e Agraria*, Portici.

1915

- Nathan Banks *A Treatise on the Acarina, Or Mites* the first comprehensive English handbook on mites.

1916

- The Japanese Beetle, Popillia japonica, was first discovered in the United States in Riverton, New Jersey during mid-August, 1916.

1920

- Alfred Kinsey became Professor of Entomology at Indiana University Bloomington.
- Ernst Jünger publishes In *Stahlgewittern*, The Storm of Steel.
- Enrico Adelelmo Brunetti *The Fauna of British*

India, Including Ceylon and Burma. Diptera 1. Brachycera published.

- Charles Thomas Brues *Insects and Human Welfare* published.
- G.D. Hale Carpenter published: *A Naturalist on Lake Victoria, with an Account of Sleeping Sickness and the Tse-tse Fly*; 1920. T.F. Unwin Ltd, London; Biodiversity Archive

1921

- Günther Enderlein Über die phyletisch älteren Stratiomyiidensubfamilien (Xylophaginae, Chiromyzinae, Solvinae, Beridinae und Coenomyiinae). *Mitt. Zool. Mus. Berl.* 10: 150-214 published.

1923

- Auguste-Henri Forel publishes a myrmecological 5-volume magnum opus, Le Monde Social des Forimis

1924

- Frederick William Frohawk's *Natural History of British Butterflies* published.

1925

- Frank M. Carpenter begins work on the Elmo Permian fossil fauna.
- Josef Fahringer *Opuscula braconolocica* (4 parts, finished 1937) begun.

1927

- Ronald A. Senior-White and Robert Knowles (entomologist) *Malaria: Its Investigation and Control, with Special Reference to Indian Conditions*. Calcutta: Thacker, Spink and Co published.
- José María Hugo de la Fuente Morales Tablas analíticas para la clasificación de los coleópteros de la Península Ibérica. Barcelona Imprenta Altés, published.

- Zeno Payne Metcalf commenced (as overall editor and author of the Homoptera sections. *General catalogue of the Hemiptera.* Completed 1971.

1928

- Jan Noskiewicz with G. Poluszynski Embryologische Untersuchungen an Strepsipteren. I. Teil: Embryogenesis der Gattung Stylops Kirby. *Akad. Umiejetnosci..*
- Leopold III of Belgium a keen amateur entomologist collects in the Dutch East Indies (1929-1929)
- Guido Grandi founded the Institute of Entomology at the University of Bologna (l'Istituto di Entomologia dell'Università di Bologna).
- World's oldest known insect, *Rhyniognatha hirsti* named by Robert John Tillyard.
- Alexander Kirilow Drenowski The Lepidoptera fauna on the high mountains of Bulgaria *Sbornik bulg. Akad. Nauk.* 23: 1-120, 1 map, published.

1930

- Camillo Acqua *Il bombice del Gelso: Nello stato normale e patologico nella tecnica dell' allevamento e della riproduzione. (Industria della preparazione del seme Bachi)*-Enc. tela. Casa Ed. di Giuseppe Cesari, published. This was an important contribution to the literature on sericulture.

1931

- Georg Hermann Alexander Ochs publishes *Über die Gyriniden-Ausbeute der Deutschen Limnologischen Sunda-Expedition mit einer Übersicht über die Gyriniden-Fauna Javas und Larvenbeschreibungen.*
- Shonen Matsumura 6,000 illustrated Insects of Japan-Empire.

1932

- A Practical Handbook of British Beetles by Norman H. Joy published by Witherby.

- Alfred Balachowsky *Étude biologique des coccides du bassin occidental de la Méditerranée* published in Paris by Lechevalier and Fils.

1934

- René Malaise invents the Malaise trap.
- Vincent Brian Wigglesworth, the "Father" of Insect Physiology, wrote the first book on insect phsyiology, *The Principles of Insect Physiology.*
- Antoni W³adys³aw Jakubski *Czerwiec polski* "Polish cochineal". Monograph on the Polish cochineal.

1935

- Gerhard Schrader discovers the powerful insecticides called organophosphates
- Walter Rothschild gives his insect collection, one of the world's largest collections of Lepidoptera to the Natural History Museum.

1936

- The Natural History Museum, London acquires the James John Joicey collection of Lepidoptera.

1938

- Lucien Chopard *La biologie des orthoptères* (Paul Lechevalier, Paris).
- Ângelo Moreira da Costa Lima commenced *Insetos do Brasil, v. 1-11. Completed* 1960.

1940

- Vladimir Nabokov begins organizing the butterfly collection of the Museum of Comparative Zoology at Harvard University.
- Ruggero Verity commenced *Farfalle,in English Butterflies, of Italia* (five volumes, 1940–1953).
- René Jeannel *Faune cavernicole de la France*, in English The Fauna of the Caves of France, published.

1941

- Zoltán Szilády A magyar birodalom legyeinek

szinopszisa. VI. Talpaslegyek, Clythidae (Platypezidae); VIII. Lauxaniidae [Synopsis of the flies of the Hungarian empire].

1942

- Woodhouse, L. G. O. & George Morrison Reid Henry. *The Butterfly Fauna of Ceylon*. Government Record Office, Colombo

1943

- André Badonnel *Faune de France. Psocoptères*. Paris. Paul Lechevalier 1943.
- Leopold Fulmek *Wirtsindex der Aleyrodiden- und Cocciden- Parasiten* Entomologische Beihefte 10: 1-100.

1944

- Enrica Calabresi commits suicide in Florence.

1945

- Edmund Brisco Ford *Butterflies* published, seminal introduction to the study of butterflies and their genetics.
- Cynthia Longfield The Odonata of South Angola. *Arquivos do Museu Bocage*, 16, Lisboa.

1946

- Institut National de la Recherche Agronomique founded.

1947

- Carlo Alonza became director of the Muséum de Gênes.

1949

- Pierre-Paul Grassé ed. Traité de Zoologie Tome IX. Insectes. Paris, 1949. 1118 p.

1950

- Maynard Jack Ramsay becomes Port Entomologist on Staten Island.

- Mahadeva Subramania Mani founded the School of Entomology at Agra, India.

1951

- Work on Sterile insect technique begun by American entomologists Raymond Bushland and Edward Knipling. For their achievement, they jointly received the 1992 World Food Prize.
- Sakae Tamura *Konchû no seitai: Raika shashinshû* or Closeups on Insects. Tokyo: Seibundo-Shinkosha
- Torkel Weis-Fogh pioneered studies of insect flight with August Krogh.

1952

- Bernard Kettlewell begins research into the influence of industrial melanism on natural selection in moths.

1953

- Willi Hennig publishes *Grundzüge einer Theorie der phylogenetischen Systematik* in Berlin. This was followed by

Kritische Bemerkungen zum phylogenetischen System der Insekten in 1953 and Phylogenetic Systematics in 1966. In these works Hennig founded cladistics.

- Sydney Skaife *African Insect Life* published.
- *Catalogue illustré des lucanides du globe* in Encyclopédie Entomologique (series A 27: 1-223) by Robert Didier and Eugene Seguy published.

1954

- Grigorij Jakovlevitsch Bey-Bienko *Insecta: Orthoptera: Tettigoniidae: Phaneropterinae.*

1955

- World programme for malaria eradication begins. Finally abandoned 1969.
- Roy Albert Crowson's *The natural classification of*

the families of Coleoptera is published. This is a classic monograph.

1957

- Clodoveo Carrión Mora died in Ecuador. Mora was a leading figure entomology of 20th century entomology in South America.

1960

- Czesaw Biesanko publishes *Álbum iconográfico dos Lepidópteros coletados por Biezanko. Papilionidae.*
- Marta Grandi *Ephemeroidea. Fauna d'Italia*

1961

- Genetic code is cracked. DNA was discovered by Friedrich Miescher in 1868, recognized as the bearer of genetic information in 1943 and revealed as a double helix by Rosalind Franklin in 1952. This leads to radical revision of the higher taxonomy of the Insecta.

1964

- Morris Rockstein's edited series — 3 vols. — *The Physiology of Insecta*
- Takashi Shirozu *Butterflies of Japan Illustrated in Colour* published in Tokyo by Hokuryu-kan.

1965

- Nikolaj Sergejevitsch Borchsenius Essay on the classification of the armoured scale insects (Homoptera, Coccoidea, Diaspididae). (In Russian.) *Entomologicheskoe Obozrenye* 44: 208-214.

1966

- First international Red Lists of endangered species were published.

1967

- Richard E. Blackwelder *Taxonomy: a Text and*

Reference Book John. Wiley and Sons, New York, published.

1968

- David Allan Young *Taxonomic Study of the Cicadellinae (Homoptera: Cicadellidae)* commenced. Finished 1986.

1969

- Reg Chapman's textbook appears — *The Insects-Structure and Function*. American Elsevier, N.Y.
- International Centre of Insect Physiology and Ecology established.

1971

- Maximilian Fischer *Index of Entomophagous Insects*. Le Francois, Paris.

1973

- Karl von Frisch awarded Nobel Prize for pioneering work on insect behaviour.
- Warwick Estevam Kerr Evolution of the population structure in bees. *Genetics* 79: 73–84.

1976

- Anastase Alfieri *The Coleoptera of Egypt* published.

1981

- Robert Michael Pyle published The National Audubon Society *Field Guide to North American Butterflies*. Knopf.

1981

- CESA Centre for Entomological Studies Ankara An international private research centre and museum on Entomology established by Ahmet Omer Kocak in Ankara, Turkey.

1984

- Árpád Soós and Lazlo Papp begin editing *Catalogue of Palaearctic Diptera*. 1984 - 1992.

- Justin O. Schmidt publishes first paper on the Schmidt Sting Pain Index.

1985

- Murray S. Blum *Fundamentals of Insect Physiology*. New York: Wiley, 1985.
- Gerald A. Kerkut and L. I. Gilbert *Comprehensive Insect Physiology, Biochemistry & Pharmacology*.

1987

- Stephen Taber III *Breeding Super Bee*. Ohio: AI Root Co, 1987.

1989

- Forensic entomologist Mark Benecke joins the punk rock band "Die Blonden Burschen", The Blonde Boys. Many past entomologists were also musical.

1990

- Hölldobler B. and E. O. Wilson publish *The Ants*
- Low cost Scanning electron microscope came into general use

1991

- Naumann, I. D., P. B. Carne, J. F. Lawrence, E. S. Nielsen, J. P. Spradberry, R. W. Taylor, M. J. Whitten and M. J. Littlejohn, eds. The Insects of Australia: A Textbook for Students and Research Workers. Volume I and II. Second Edition. Carlton, Victoria, Melbourne University Press.

1993

- Edward Grumbine, *Ghost Bears: Exploring the Biodiversity Crisis* reflects growing concerns. Insects are major indicators of environmental destruction and impending mass extinction.

1994

- Hoy, M. *Insect molecular genetics*. An introduction to principles and applications.
- Vladimir Nikolayevich Beklemiwev (Methodology of systematics). KMK Scientific Press Ltd.

1995

- Yuri Petrovich Korshunov and Pavel Yunievich Gorbunov *Butterflies of the Urals, Siberia and Far East* published.

1996

- Microcosmos released in France.

1997

- Perry Adkisson receives World Food Prize for his work on Integrated Pest Management.

1998

- Paul R. Ehrlich publishes *Betrayal of Science and Reason: How Anti-Environment Rhetoric Threatens Our Future* (1998, co-authored with his wife)
- Phylocode proposed following a meeting at Harvard University.

1999

- Ebbe Schmidt Nielsen instrumental in setting up the Global Biodiversity Information Facility

2000

- Loïc Matile *Diptères d'Europe Occidentale* Tomes 1 and 2 *Atlas d'Entomologie*. Editions N. Boubée, Paris.

2001

- First volume of American Beetles published. Ross H. Arnett, Jr. and Michael C. Thomas.

2002

- Alex Rasnitsyn with D.L.J. Quicke *History of Insects*. Kluwer Academic Publishers.

2004

- Gilbert, L.I. (ed.). 2004. *Comprehensive molecular insect science*, 7 vols. Elsevier Pergamon, published in St. Louis
- A paper in *Science* found that *Culex pipiens* mosquitoes existed in two populations in Europe,

one which bites birds and one which bites humans. In North America 40% of *Culex pipiens* were found to be hybrids of the two types which bite both birds and humans, providing a vector for West Nile virus. This is thought to provide an explanation of why the West Nile disease has spread more quickly in North America than Europe.

2005

- The Insect Biocontrol Laboratory at the Henry A. Wallace Beltsville Agricultural Research Center in the United States develops DNA fingerprinting tools that match hard-to-identify larvae to adults that have been positively identified.
- Michael S. Engel and David Grimaldi *Evolution of the Insects* published.

FORENSIC ENTOMOLOGY

Forensic entomology is the application and study of insect and other arthropod biology to criminal matters. Forensic entomology is primarily associated with death investigations; however, it may also be used to detect drugs and poisons, determine the location of an incident, and find the presence and time of the infliction of wounds. Forensic entomology can be divided into three subfields: urban, stored-product and medico-legal/medico-criminal entomology.

History

Historically, there have been several accounts of applications for, and experimentation with, forensic entomology. The concept of forensic entomology dates back to at least the 14th century. However, only in the last 30 years has forensic entomology been systematically explored as a feasible source for evidence in criminal investigations. Through their own experiments and interest in arthropods and death, Song Ci, Francesco Redi, Bergeret d'Arbois, Jean Pierre Mégnin and the German doctor Hermann

Reinhard have helped to lay the foundations for today's modern forensic entomology.

Song Ci

Song Ci (also known as Sung Tz'u) was a lawyer and death investigator who lived in China in the late 13th century. In 1247 AD Song Ci wrote a book entitled *Washing Away of Wrongs*. In this book Song Ci depicts several cases in which he took notes on how a person died and elaborates on probable causes. He goes into detail on how to examine a corpse both before and after burial. He also explains the process of how to determine a probable cause of death. The purpose of this book was to be used as a guide for other investigators so they could assess the scene of the crime effectively. His level of detail in explaining what he observed in all his cases laid down the fundamentals for modern forensic entomologists and is the first recorded account in history of someone using forensic entomology for judicial means. This book was immensely popular and represented the first time that the general public became aware that insects could be used in criminal investigations.

Francesco Redi

In 1668, Italian physician Francesco Redi disproved the theory of spontaneous generation. The accepted theory of Redi's day claimed that maggots developed spontaneously from rotting meat. In an experiment, he used samples of rotting meat that were either fully exposed to the air, partially exposed to the air, or not exposed to air at all. Redi showed that both fully and partially exposed rotting meat developed fly maggots, whereas rotting meat that was not exposed to air did not develop maggots. This discovery completely changed the way people viewed the decomposition of organisms and prompted further investigations into insect life cycles and into entomology in general.

Bergeret d'Arbois

Dr Louis François Etienne Bergeret (1814–1893) was

a French hospital physician, and was the first to apply forensic entomology to a case. In a case report published in 1855 he stated a general life cycle for insects and made many assumptions about their mating habits. Nevertheless these assumptions led him to the first application of forensic entomology in an estimation of post-mortem interval (PMI). His report used forensic entomology as tool to prove his hypothesis on how and when the person had died.

Hermann Reinhard

The first systematic study in forensic entomology was conducted in 1881 by Hermann Reinhard, a German medical doctor who played a vital role in the history of forensic entomology. He exhumed many bodies and demonstrated that the development of many different types of insect species could be tied to buried bodies. Reinhard conducted his first study in east Germany, and collected many Phorid flies from this initial study. He also concluded that the development of only some of the insects living with corpses underground were associated with them, since there were 15-year-old beetles who had little direct contact with them. Reinhard's works and studies were used extensively in further forensic entomology studies.

Jean Pierre Mégnin

Jean Pierre Mégnin (1828–1905), an army veterinarian, published many articles and books on various subjects including the books *Faune des Tombeaux* and *La Faune des Cadavres*, which are considered to be among the most important forensic entomology books in history. In his second book he did revolutionary work on the theory of predictable waves, or successions of insects onto corpses. By counting numbers of live and dead mites that developed every 15 days and comparing this with his initial count on the infant, he was able to estimate how long that infant was dead. In this book he asserted that exposed corpses were subject to eight successional waves, whereas buried corpses were only subject to two waves. Mégnin made many great discoveries that helped shed new light on many of the general characteristics of decaying flora and fauna. Mégnin's work and study of the larval and adult forms of insect families found in cadavers sparked the interest of future entomologists and encouraged more research in the link between arthropods and the deceased, and thereby helped to establish the scientific discipline of forensic entomology.

Forensic Entomology Subfields

Urban Forensic Entomology

Urban forensic entomology typically concerns pests infestations in buildings gardens or that may be the basis of litigation between private parties and service providers such as landlords or exterminators. Urban forensic entomology studies may also indicate the appropriateness of certain pesticide treatments and may also be used in stored products cases where it can help to determine chain of custody, when all points of possible infestation are examined in order to determine who is at fault.

Stored-Product Forensic Entomology

Stored-product forensic entomology is often used in litigation over infestation or contamination of commercially distributed foods by insects.

Medico-Legal Forensic Entomology

Medicolegal forensic entomology covers evidence that may be gathered through arthropod studies at events such as murder, suicide, rape, physical abuse and contraband trafficking. In murder investigations it deals with which insects eggs when and where, and in what order they appear in dead bodies. This can be helpful in determining a post mortem interval (PMI) and location of a death in question. Since many insects exhibit a degree of endemism (occurring only in certain places), or have a well-defined phenology (active only at a certain season, or time of day), their presence in association with other evidence can demonstrate potential links to times and locations where other events may have occurred). Another area covered by medicolegal forensic entomology is the relatively new field of entomotoxicology. This particular branch involves the utilization of entomological specimens found at a scene in order to test for different drugs that may have possibly played a role in the death of the victim.

Insect Types

There are many different types of insect studied in forensic entomology. The insects listed below are mostly necrophagous (corpse-eating) and are particularly relevant to medicolegal entomological investigations. This is not a full list, as there are many variations due to climate. The order in which insects feed on a corpse is known as faunal succession.

Flies

Flies (order diptera) are often first on the scene. They prefer a moist corpse for their offspring (maggots) to feed on. The most significant types of fly include:

- Blow flies – Family Calliphoridae- Flies in this family are often metallic in appearance and between ten to 12 mm in length. In addition to the name blow-fly, some members of this family are known as blue bottle fly, cluster fly, greenbottles, or black blowfly. A characteristic of the blow-fly is its 3-segmented antennae. Hatching from an egg to the first larval stage takes from eight hours to one day. Larvae have three stages of development (called instars); each stage is separated by a molting event. Worldwide, there are 1100 known species of blowflies, with 228 species in the Neotropics, and a large number of species in Africa and Southern Europe. The most common area to find Calliphoridae species are in the countries of India, Japan, Central America, and in the southern United States. The typical habitat for blow-flies are temperate to tropical areas that provide a layer of loose, damp soil and litter where larvae may thrive and pupate. The forensic importance of this fly is that it is the first insect to come in contact with carrion because they have the ability to smell death from up to ten miles (16 km) away.

- Flesh flies – Family Sarcophagidae- Most flesh flies breed in carrion, dung, or decaying material, but a few species lay their eggs in the open wounds of mammals; hence their common name. Characteristics of the flesh-fly is its 3-segmented antennae. They are medium-sized flies with black and gray longitudinal stripes on the thorax and checkering on the abdomen. Flesh-flies, being viviparous, frequently give birth to live young on corpses of human and other animals, at any stage of decomposition, from newly dead through to bloated or decaying (though the latter is more common).
- House fly – Family Muscidae- is the most common of all flies found in homes, and indeed one of the most widely distributed insects; it is often considered a pest that can carry serious diseases. The adults are 6–9 mm long. Their thorax is gray, with four longitudinal dark lines on the back. The underside of their abdomen is yellow, and their whole body is covered with hair. Each female fly can lay up to 500 eggs in several batches of about 75 to 150 eggs. Genus Hydrotaea are of particular forensic importance.
- Cheese flies – Family Piophilidae - Most are scavengers in animal products and fungi. The best-known member of the family is Piophila casei. It is a small fly, about four mm (1/6 inch) long, found worldwide. This fly's larva infests cured meats, smoked fish, cheeses, and decaying animals and is sometimes called the cheese skipper for its leaping ability. Forensic entomology uses the presence of Piophila casei larvae to help estimate the date of

death for human remains. They do not take up residence in a corpse until three to six months after death. The adult fly's body is black, blue-black, or bronze, with some yellow on the head, antennae, and legs. The wings are faintly iridescent and lie flat upon the fly's abdomen when at rest. At four mm (1/6 inch) long, the fly is one-third to one-half as long as the common housefly.

- Coffin flies – Phoridae
- Lesser corpse flies – Sphaeroceridae
- Lesser house flies – Fanniidae
- Black scavenger flies – Sepsidae
- Sun flies - Heleomyzidae
- Black soldier fly - Stratiomyidae - have potential for use in forensic entomology. The larvae are common scavengers in compost heaps, are found in association with carrion, can be destructive pests in honey bee hives, and are used in manure management (for both house fly control and reduction in manure volume). The larvae range in size from 1/8 to 3/4 of an inch (3 to 19 millimeters). The adult fly is a mimic, very close in size, color, and appearance to the organ pipe mud dauber wasp and its relatives.

Beetles

Beetles (Order Coleoptera) are generally found on the corpse when it is more decomposed. In drier conditions, the beetles can be replaced by moth flies (Psychodidae).

- Rove beetles – Family Staphylinidae – are elongate beetles with small elytra (wing covers) and large jaws. Like other beetles inhabiting carrion, they have fast larval development with only three larval stages. Creophilus species are common predators of carrion, and since they are large, are a very visible component of the fauna of corpses. Some adult Staphylinidae are early visitors to a corpse, feeding on larvae of all species of fly, including the later predatory fly larvae. They lay their eggs in the corpse, and the emerging larvae are also predators. Some species have a long development time in the egg, and are common only during the later stages of decomposition. Staphylinids can also tear open the pupal cases of flies, to sustain themselves at a corpse for long periods.

- Hister beetles – Family Histeridae. Adult histerids are usually shiny beetles (black or metallic-green) which have an introverted head. The carrion-feeding species only become active at night when they enter the maggot-infested part of the corpse to capture and devour their maggot prey. During daylight they hide under the corpse unless it is sufficiently decayed to enable them to hide inside it. They have fast larval development with only two larval stages. Among the first beetles to arrive at a corpse are Histeridae of the genus *Saprinus*. *Saprinus* adults feed on both the larvae and pupae of blowflies, although some have a preference for fresh pupae. The adults lay their eggs in the corpse, inhabiting it in the later stages of decay.
- Carrion beetles – Family Silphidae- Adult Silphidae have an average size of about 12 mm. They are also referred to as burying beetles because they dig and bury small carcasses underground. Both parents tend to their young and exhibit communial breeding. The male carrion beetle's job in care is to provide protection for the breed and carcass from competitors.
- Ham beetles – Family Cleridae
- Carcass beetles – Family Trogidae
- Skin/hide beetles – Family Dermestidae. Hide beetles are important in the final stages of decomposition of a carcass. The adults and larvae feed on the dried skin, tendons and bone left by fly larvae. Hide beetles are the only beetle with the

enzymes necessary for breaking down keratin, a protein component of hair.

- Scarab beetles – Family Scarabaeidae- Scarab beetles may be any one of around 30,000 beetle species worldwide that are compact, heavy-bodied and oval in shape. The flattened plates, which each antenna terminates, are fitted together to form a club. The outer edges of the front legs may also be toothed or scalloped. Scarab beetles range from 0.2 to 4.8 in (5.1 to 120 mm) in length. These species are known for being one of the heaviest insect species.
- Sap beetles – Family Nitidulidae

Mites

Many mites (Class Acari) feed on corpses with *Macrocheles* mites common in the early stages of decomposition, while Tyroglyphidae and Oribatidae mites such as Rostrozetes feed on dry skin in the later stages of decomposition.

Nicrophorus beetles often carry on their bodies the mite *Poecilochirus* which feed on fly eggs. If they arrive at the corpse before any fly eggs hatch into maggots, the first eggs are eaten and maggot development is delayed. This may lead to incorrect PMI estimates. *Nicrophorus* beetles find the ammonia excretions of blowfly maggots toxic, and the *Poecilochirus* mites, by keeping the maggot population low, allow *Nicrophorus* to occupy the corpse.

Moths

Moths (Order Lepidoptera) specifically clothes-moths – Family Tineidae – are closely related to butterflies. Most species of moth are nocturnal, but there are crepuscular and diurnal species. Moths feed on mammalian hair during their larval stages and may forage on any hair that remains on a body. They are amongst the final animals contributing to the decomposition of a corpse.

Wasps, Ants and Bees

Wasps, ants, and bees (Order Hymenoptera) are not necessarily necrophagous. While some feed on the body, some are also predatory, and eat the insects feeding on the body. Bees and wasps have been seen feeding on the body during the early stages. This may cause problems for murder cases in which larval flies are used to estimate the post mortem interval since eggs and larvae on the body may have been consumed prior to the arrival on scene of investigators.

- *Wasps* – (particularly family Vespidae). Wasps exhibit a range of social difficulty, from private living to eusocial colonies. The non-breeding creature cares for the young or defend and supply for the group. Wasps are commentable for studies of evolutionary origin and maintenance of social behaviour in animals.
- *Ants* – Family Formicidae. Among the most widespread and damaging of introduced species are ants. Many ants share some characteristics that ease their preamble, institution, and subsequent range expansion. One feature of their importance is the ability to establish numerically large, ecologically dominant colonies.
- *Bees* – Superfamily Apoidea. Forensic entomologists have used bees in several cases where parents have used bees to sting their children as a form of discipline. Also, entomologists have been called upon to determine whether or not bees or wasps have been the cause of an accident. Whether through their presence or by stinging it has be speculated that these insects have been the cause of numerous automobile accidents.

Modern Techniques

Many new techniques have been discovered and used in order to more accurately gather evidence, or possibly introduce an entire new way to look at old information. Over the years it has become more popular as case studies open doors to new ideas and techniques, once though defunct, have since proven to be invaluable in some courtroom battles. Forensic entomology not only uses arthropod biology, but it pulls from other sciences, introducing fields like chemistry and genetics, exploiting their inherent synergy through the use of DNA in forensic entomology.

Scanning Electron Microscopy

Usually fly larvae are used to aid in the determination of a *PMI*. However, sometimes the body may not contain maggots and only the eggs are present. In order for the data to be useful the eggs must be identified down to a species level to get an accurate estimate for the PMI. There are many techniques currently being developed to differentiate between the various species of forensically important insects. A study in 2007 demonstrates a technique that can use scanning electron microscopy (*SEM*) to identify key morphological features of eggs and maggots. Some of the morphological differences that can help identify the different species are the presence/absence of anastomosis, the presence/absence of holes, and the shape and length of the median area. The SEM method provides an array of morphological features for use in identifying fly eggs; however, this method does have some disadvantages. The main disadvantage is that it requires expensive equipment and can take time to identify the species from which the egg originated, so it may not be useful in a field study or to quickly identify a particular egg. The SEM method is good if ample time and resources to determine the species of the particular fly egg are plentiful. The ability to use these morphological differences gives forensic entomologists a powerful tool that can help with estimating a post mortem interval, along with other relevant information.

Potassium Permanganate Staining

Sometimes scanning electron microscopy is not available

and a quicker and lower cost technique can be found in potassium permanganate staining. This process involves a few basic steps. Once the eggs are collected, they are rinsed with a normal saline solution and then moved to a glass petri dish. The eggs are then soaked in a 1% potassium permanganate solution for one minute. Then the eggs are dehydrated and mounted onto a slide for observation. These slides can be used with any light microscope with a calibrated eyepiece to compare various morphological features. The most important and useful features observed for identifying eggs are things like the size, length, and width of the plastron, as well as the morphology of the plastron in the area around the micropyle. The various measurements and observations are then compared to standards for forensically important species and used to determine the species of the egg.

Mitochondrial DNA

In 2001, a method was devised by Jeffrey Wells and Felix Sperling to use mitochondrial DNA to differentiate between different species of the subfamily Chrysomyinae. This is particularly useful when working on determining the identity of specimens that do not have distinctive morphological characteristics at certain life stages.

Mock Crime Scenes

A valuable tool that is becoming very common in the training of forensic entomologists is the use of mock crime scenes using pig carcasses. The pig carcass represents a human body and can be used to illustrate various environmental effects on both arthropod succession and the estimate of the post mortem interval.

Gene Expression Studies

Although physical characteristics and sizes at various instars have been used to estimate fly age, more recently a study has been conducted to determine the age of an egg based on the expression of particular genes. This is particularly useful in developmental stages that do not change in size, such as the egg or pupa, where only a general time interval can be estimated based on the duration of the particular developmental stage. This is done by breaking the stages down into smaller units separated by predictable changed in gene expression. Three genes were measured in an experiment with *Drosophila melanogaster*: bicoid (bcd), slalom (sll), and chitin synthase (cs). These three genes were used because they are likely to be in varied levels during different times of the egg development process. These genes all share a linear relationship in regards to age of the egg; that is, the older the egg is the more of the particular gene is expressed. However, all of the genes are expressed in varying amounts. Different genes on different loci would need to be selected for another fly species. The genes expressions are mapped in a control sample to formulate a developmental chart of the gene expression at certain time intervals.

This chart can then be compared to the measured values of gene expression to accurately predict the age of an egg to within two hours with a high confidence level. Even though this technique can be used to estimate the age of an egg, the feasibility and legal acceptance of this must be considered for it to be a widely utilized forensic technique. One benefit of this would be that it is like other DNA-based techniques so most labs would be equipped to conduct similar experiments without requiring new capital investment. This style of age determination is in the process of being used to more accurately find the age of the instars and pupa; however, it is much more complicated, as there are more genes being expressed during these stages. The hope is that through this, and other techniques similar to it, a more accurate PMI can be obtained.

Insect Activity Case Study

A preliminary investigation of insect colonization and succession on remains in New Zealand revealed the following results on decay and insect colonization.

Open Field Habitat

This environment had a daily average maximum temperature of 19.4 degrees Celsius and a daily minimum temperature of 11.1 degrees Celsius. The average rainfall for the first 3 weeks in this environment was 3.0 mm/day. Around days 17–45, the body began to start active decay. During this stage, the insect successions started with Calliphora stygia, which lasted until day 27.

The larvae of Chrysomya rufifacies were present between the day 13 and day 47. The H. rostrata, larvae of Lucilia sericata, Psychodidae family, and sylvicola were found to occur relatively late in the body's decay.

Coastal Sand-Dune Habitat

This environment had an average daily maximum temperature of 21.4 degrees Celsius and minimum of 13.5 degrees Celsius. The daily average rainfall was recorded as 1.4 mm/day for the first 3 weeks. The post-decay time interval, beginning at day six after death and ending around day 15 after death, is greatly reduced from the average post-decay time, due to the high average temperature of this environment. Insects obtained late in the post-active stage include the Callihora quadrimaculat, adult Phaeroceridae, Psychodidae and Piophilidae (no larvae from this family were obtained in recovery).

Native Bush Habitat

This environment had recorded daily average maximum and minimum temperatures were 18.0 and 13.0 degrees Celsius, respectively. The average rainfall in this habitat was recorded at 0.4 mm/day. After the bloat stage, which lasted until day seven after death, post-active decay began around day 14. In this habitat, the H. rostrata, Phoridae adult, Sylvicola larvae and adult are the predominant species remaining on the body during the pre-skeletonization stages.

Forensic Entomology in Scientific and Fictional Literature in the 19th and Early 20th Centuries

Throughout its history the study of forensic entomology has not remained an esoteric science reserved only for entomologists and forensic scientists. Early twentieth-century popular scientific literature began to pique a broader interest in entomology. The very popular ten-volume book series, Alfred Brehem's *Thierleben* (Life of Animals, 1876–1879) expounded on many zoological topics, including arthropods. The accessible writing style of French entomologist Jean-Henri Fabre was also instrumental in the popularization of entomology. His collection of writings *Souvenirs Entomologique*, written during the last half of the 19th century, is especially useful because of the meticulous attention to detail to the observed insects' behaviours and life cycles.

The real impetus behind the modern cultural fascination with solving crime using entomological evidence can be traced back to the works *Faune de Tombeaux* (Fauna of the Tomb, 1887) and *Les Faunes des Cadavres* (Fauna of the Cadaver, 1894) by French veterinarian and entomologist Jean Pierre Mégnin. These works made the concept of the process of insect ecological succession on a corpse understandable and interesting to an ordinary reader in a way that no other previous scientific work had done. It was after the publication of Mégnin's work that the studies of forensic science and entomology became an established part of Western popular culture, which in turn inspired other scientists to continue and expand upon his research. The use of forensic science, including forensic entomology, became a popular part of fiction, beginning with the stories of Edgar Allan Poe, often thought of as the inventor of the modern detective story.. He is noted for including graphic depictions of human decomposition in some of his works, including entomological references. The poem "The Conqueror Worm", published in 1843, features death symbolized through a "worm" feasting upon human remains. The worm is a reference to the maggots which are present on a corpse after death and aid in decomposition. In his short story "The Premature Burial," published in 1850, he discusses the Victorian Era's fascination with being buried alive, including references to autopsy procedures of the day and the known pattern of human decomposition.

In Film

Silence of the Lambs

In the movie *The Silence of the Lambs*, a series of murders are committed in which the killer skins his victims, leading the investigators to dub the unidentified killer "Buffalo Bill". In a scene depicting an autopsy of one of the victims, the investigators notice something in the victim's throat. The pupa of a Death's-head Hawkmoth is removed with forceps. In another scene, entomologists in the film explain to investigators that the particular species discovered on the victim, identified as *Acherontia styx*, is not native to the United States and must have been imported by the killer. Because of the enormous popularity of the film, these scenes referencing the overlap between entomology and forensic science are what most people associate with the work of forensic entomologists.

My Girl

In the 1991 movie *My Girl*, protagonist Thomas Jay dies due to anaphylaxis after being stung by bees, to which he was mortally allergic. Often with such cases death due to anaphylaxis, an immediate cause of death may not be apparent, requiring the expertise of a forensic entomologist.

TV Shows

CSI: Crime Scene Investigation: In the television show *CSI: Crime Scene Investigation*, forensic entomologist Gil Grissom is famous for his knowledge of bugs. He is a valuable part of the CSI team with his ability to distinguish insect types and their significance at the crime scene.

Bones: In the television show *Bones*, Jack Hodgins is the forensic team's entomologist, frequently working to identify the postmortem interval from insects found with the remains.

FORENSIC ENTOMOLOGICAL DECOMPOSITION

Medicolegal entomology is a branch of forensic entomology that applies the study of insects to criminal investigations, and is commonly used in death investigations for estimating the post-mortem interval (PMI). One method of obtaining this estimate uses the time and pattern of arthropod colonization. This method will provide an estimation of the period of insect activity, which may or may not correlate exactly with the time of death. While insect successional data may not provide as accurate an estimate during the early stages of decomposition as developmental data, it is applicable for later decompositional stages and can be accurate for periods up to a few years.

Decomposition

Decomposition is a continuous process that is commonly divided into stages for convenience of discussion. When studying decomposition from an entomological point of view and for the purpose of applying data to human death investigations, the domestic pig Sus scrofa (Linnaeus) is considered to be the preferred human analogs. In entomological studies, five stages of decomposition are commonly described: (1) Fresh, (2) Bloat, (3) Active Decay, (4) Advanced or Post-Decay, and (5) Dry Remains. While the pattern of arthropod colonization follows a reasonably predictable sequence, the limits of each stage of decomposition will not necessarily coincide with a major change in the faunal community. Therefore, the stages of decomposition are defined by the observable physical changes to the state of the carcass. A pattern of insect succession results as different carrion insects are attracted to the varying biological, chemical and physical changes a carcass undergoes throughout the process of decay.

A decaying carcass provides "a temporarily, rapidly

changing resource which supports a large, dynamic arthropod community."

—*M. Grassberger and C. Frank*

Fresh Stage

The fresh stage of decomposition is generally described as the period between the moment of death and when the first signs of bloat are apparent. There are no outward signs of physical change, though internal bacteria have begun to digest organ tissues. No odor is associated with the carcass. Early post-mortem changes, used by pathologists as medical markers for early post-mortem interval estimations, have been described by Goff and include livor mortis, rigor mortis and algor mortis. The first insects to arrive at decomposing remains are usually Calliphoridae, commonly referred to as blow flies. These flies have been reported to arrive within minutes of death or exposure, and deposit eggs within 1-3 hours. Adult flies of the families Sarcophagidae (flesh flies) and Muscidae are also common in this first stage of decomposition. First eggs are laid in or near the natural orifices of the head and anus, as well as at the site of perimortem wounds. Depending on the rate of decomposition and the development time of particular blowfly species, eggs may hatch and young larvae begin to feed on tissues and liquids while the carcass is still classified in the fresh stage. Adult ants may also be seen at a carcass during the fresh stage. Ants will feed both on the carcass flesh as well as eggs and young larvae of first arriving flies.

Bloat Stage

The first visible sign of the bloat stage is a slight inflation of the abdomen and some blood bubbles at the nose. Activity of anaerobic bacteria in the abdomen create gases, which accumulate and results in abdominal bloating. A colour change is observed in the carcass flesh, along with the appearance of marbling. During the bloat stage the odor of putrefaction becomes noticeable. Blowflies remain present in great numbers during the bloat stage, and blowflies, flesh flies and muscids continue to lay eggs. Insects of the families Piophilidae and Fanniidae arrive during the bloat stage. Ants continue to feed on the eggs and young larvae of flies. The first species of Coleoptera arrive during the bloat stage of decomposition, including members of the families Staphylinidae (rove beetles), Silphidae (carrion beetles) and Cleridae. These beetles are observed feeding on fly eggs and larvae. Beetle species from the families Histeridae may also be collected during this stage, and are often hidden beneath remains.

Active Decay Stage

The beginning of active decay stage is marked by the deflation of the carcass as feeding Dipteran larvae pierce the skin and internal gases are released. During this stage the carcass has a characteristic wet appearance due to the liquefaction of tissues. Flesh from the head and around the anus and umbilical cord is removed by larval feeding activity. A strong odor of putrefaction is associated with the carcass. Feeding larvae of Calliphoridae flies are the dominant insect group at carcasses during the active decay stage. At the beginning of the stage larvae are concentrated in natural orifices, which offer the least resistance to feeding. Towards later stages, when flesh has been removes from the head and orifices, larvae become more concentrated in the thoracic and abdominal cavities. Adult calliphorids and muscids decreased in numbers during this stage, and were not observed to be mating. However, non-Calliphoridae Dipterans are collected from carcasses. The first members of Sepsidae arrive at the carcass during the active decay stage. Members of Coleoptera become the dominant adult insects at the site of remains. In particular, the numbers of staphylinids and histerids increase.

Advanced Decay Stage

Most of the flesh is removed from the carcass during the advanced decay stage, though some flesh may remain in the abdominal cavity. Strong odors of decomposition begin to fade. This stage marks the first mass migration of third instar calliphorid larvae from the carcass Piophilidae larvae may also be collected at this stage. Few adult calliphoridae are attracted to carcasses in advanced decay. Adult Dermestidae (skin beetles) arrive at the carcass; adult dermestid beetles may be common, whereas larval stages are not

Dry Decay

The final stage of decomposition is dry remains. Payne described a total of six stages of decay, the last two being separate dry and remains. As these stages are nearly impossible to distinguish between, many entomological studies combine the two into a single final stage. Very little remains of the carcass in this stage, mainly bones, cartilage and small bits of dried skin. There is little to no odor associated with remains. Any odor present may range from that of dried skin to wet fur. The greatest number of species are reported to occur in the late decay and dry stages. The dry decay stage is characterized by the movement from previously dominant carrion fauna to new species. Very few adult callihorids are attracted to the carcasss at this stage, and adult piophilids emerge. The dermestid beetles, common in advanced decay, leave the carcass. Non-carrion insects that commonly arrive at remains in dry decay are centipedes, millipedes, isopods, snails and cockroaches.

Factors Affecting Decomposition

Understanding how a corpse decomposes and the factors that may alter the rate of decay is extremely important for evidence in death investigations. Campobasso, Vella, and Introna consider the factors that may inhibit or favour the colonization of insects to be vitally important when determining the time of insect colonization.

Temperature and Climate

Low temperatures generally slow down the activity of

blow-flies and their colonization of a body. Higher temperatures in the summer favour large maggot masses on the carrion. Dry and windy environments can dehydrate a corpse, leading to mummification. Dryness causes cessation in bacterial growth since there are no nutrients present to feed on.

Access

Access to the body can limit which insects can get to the body in order to feed and lay eggs. In the United States, corpses found in brightly lit areas are generally inhabited by *Lucilia illustris*. This is in contrast to *Phormia regina*, which prefers more shaded areas. Darkness, cold, and rain limit the amount of insects that would otherwise colonize the body. A submerged corpse can vary in temperature and is colonized by very few terrestrial insects. Fish, crustaceans, aquatic insects and bacteria would be the likely fauna in this case. Bodies that have been buried are harder to get to than freely available bodies which limits the availability of certain insects to colonize. The Coffin fly *Megaselia scalaris* is one of the few fly species seen on buried bodies because it has the ability to dig up to six feet underground to reach a body and oviposit.

Reduction and Cause of Death

Scavengers and carnivores such as wolves, dogs, cats, beetles, and other insects feeding on the remains of a carcass can make determining the time of insect colonization much harder. This is because the decomposition process has been interrupted by factors that may speed up decomposition. Corpses with open wounds, whether pre or post mortem, tend to decompose faster due to easier insect access. The cause of death likewise can leave openings in the body that allow insects and bacteria access to the inside body cavities in earlier stages of decay. Flies oviposit eggs inside natural openings and wounds that may become exaggerated when the eggs hatch and the larvae begin feeding.

Clothing and Pesticides

Wraps, garments, and clothing have shown to effect the rate of decomposition because the corpse is covered by some type of barrier. Wraps, such as tight fighting tarps can advance the stages of decay during warm weather when the body is outside. However, loose fitting coverings that are open on the ends may aid colonization of certain insect species and keep the insects protected from the outside environment. This boost in colonization can lead to faster decomposition. Clothing also provides a protective barrier between the body and insects that can delay stages of decomposition. For instance, if a corpse is wearing a heavy jacket, this can slow down decomposition in that particular area and insects will colonize elsewhere. Bodies that are covered in pesticides or in an area surrounded in pesticides may be slow to have insect colonization. The absence of insects feeding on the body would slow down the rate of decomposition.

Percent Body fat of Corpse

More fat on the body allows for faster decomposition. This is due to the composition of fat, which is high in water content. Larger corpses with higher percent body fat also tend to retain heat much longer than corpses with less body fat. Higher temperatures favour the reproduction of bacteria inside high nutrient areas of the liver and other organs.

Drugs

On occasion, drugs that are present in the body at death can also affect how fast insects break down the corpse. Development of these insects can be sped up by cocaine and slowed down by drugs containing arsenic.

Current Research

New research in the related field entomotoxicology is currently studying the effects of drugs on the development of insects who have fed on the decomposing tissue of a drug user. The effects of drugs and toxins on insect development are proving to be an important factor when determining the insect colonization time. It has been shown that cocaine use can accelerate the development of maggots. In one case, *Lucilia sericata* larvae that fed in the nasal cavity of a cocaine abuser, grew over 8 mm longer than larvae of the same generation found elsewhere on the body. Other researchers in entomotoxicology are developing techniques to detect and measure drug levels in older fly pupae. This research is useful for determining cause of death for bodies that are found during later stages of decay. To this date, bromazepam, levomepromazine, malathion, phenobarbital, trazolam, oxazepam, alimemazine, clomipramine, morphine, mercury, and copper have been recovered from maggots.

Understanding the stages of decomposition, the colonization of insects, and factors that may affect decomposition and colonization are key in determining forensically important information about the body. Different insects colonize the body throughout the stage of decomposition. In entomological studies these stages are commonly described as fresh, bloat, active decay, advanced decay and dry decay. Studies have shown that each stage is characterized by particular insect species, the succession of which is depends on chemical and physical properties of remains, rate of decomposition and environmental factors. Insects associated with decomposing remains may be useful in determining post-mortem interval, manner of death, and the association of suspects. Insect species and their times of colonization will vary according to the geographic region, and therefore may help determine if remains have been moved.

ECONOMIC ENTOMOLOGY

Economic entomology is a field of entomology, which involves the study of insects that are of benefit or those that cause harm to humans, domestic animals, and crops.

Insects that cause losses are termed as pests. Some species can cause indirect damage by spreading diseases and these are termed as vectors. Those that are beneficial include those reared for food such as honey, substances such as lac or pigments and for their role in pollinating crops and controlling pests.

Early Workers

Economic entomology began in China where in 1061 Shen Kuo described the role of predatory insects in protecting crops from insect pests and ....*to be continued* In the 18th century many works were published on agriculture. Many contained accounts of pest insects. In France Claude Sionnest (1749–1820) was a notable figure.

Early Literature

19th Century

The most able exponent of this subject in Great Britain was John Curtis, whose treatise *Farm Insects*, published in 1860, was once the standard British work dealing with the insect pests of corn, roots, grass and stored corn. The most important works dealing with fruit and other pests were by Saunders, Joseph Albert Lintner, Charles Valentine Riley, Mark Vernon Slingerland and others in America and Canada. In Europe the earliest works were by Ernst Ludwig Taschenberg, Sven Lampa (1839–1914), Enzio Rafael Reuter (1867–1951) and Vincenze Kollar. Charles French (1842–1933), Walter Wilson Froggatt (1858–1937) and Henry Tryon (1856–1943) pioneered in Australia. It was not until the last quarter of the 19th century that any real advance was made in the study of economic entomology. Among the early writings, besides the book of Curtis, there was also a publication by Pohl and Kollar, entitled *Insects Injurious to Gardeners, Foresters and Farmers*, published in 1837, and Taschenberg's *Praktische Insecktenkunde*. During the 19th century Italian entomologists made significant progress in controlling diseases of the Silk moth which supported the silk industry, in the control of agricultural pests and in stored product entomology. Significant figures were: Agostino Bassi (1773–1856), Camillo Róndani (1808–1879), Adolfo Targioni Tozzetti (1823–1902), Pietro Stefanelli (1835, 1919), Camillo Acqua (1863–1936) Antonio Berlese (1863–1927), Gustavo Leonardi (1869–1918) and Enrico Verson (1845–1927). In France Etienne Laurent Joseph Hippolyte Boyer de Fonscolombe, Charles Jean-Baptiste Amyot, Émile Blanchard, Valéry Mayet and Claude Charles Goureau were early workers as was Jean Victoire Audouin the author of *Histoire des insectes nuisibles à la vigne et particulièrement de la Pyrale*, Philippe Alexandre Jules Künckel d'Herculais and Jean-Étienne Girard. American literature began as far back as 1788, when a report on the Hessian fly was issued by Sir Joseph Banks; in 1817 Thomas Say began his writings; while in 1856 Asa Fitch started his report on *Noxious Insects of New York*. Also in America, Matthew Cooke wrote *Treatise on the Insects Injurious to Fruit and Fruit Trees of the State of California, and Remedies Recommended for Their Extermination*, published in 1881. The Englishman Frederick Vincent Theobald wrote A text-book of agricultural zoology in 1890. It became a standard text worldwide.

20th Century

Among the most important reports early in the 20th century were those of Charles Valentine Riley, published by the U.S. Department of Agriculture, extending from 1878 to his death, in which is embodied an enormous amount of valuable material. At his death the work fell to Professor Leland Ossian Howard, in the form of *Bulletin of the U.S. Department of Agriculture*. The chief writings of J. A. Lintner extend from 1882 to 1898, in yearly parts, under the title of *Reports on the Injurious Insects of the State of New York*. Another significant contributor to the entomological literature of the United states was Charles W. Woodworth. The Florida entomologist Wilmon Newell was a pioneer of pest control as was Clarence Preston Gillette. In India Thomas Bainbrigge Fletcher, who succeeded Harold Maxwell-Lefroy and Lionel de Nicéville as the first Imperial Entomologist, wrote *Some South Indian insects and other animals of importance considered especially from an economic point of view*, an influential work in the subcontinent. In France Alfred Balachowsky was a key figure. In the last quarter of the 20th century new techniques were pioneered and new theories developed, for instance Integrated Pest Management by Ray F. Smith.

Harmful Insects

Insects considered *pests* of some sort occur among all major living orders with the exception of Ephemeroptera (mayflies), Odonata, Plecoptera (stoneflies), Embioptera (webspinners), Trichoptera (caddisflies), Neuroptera (in the broad sense), and Mecoptera (also, the tiny groups Zoraptera, Grylloblattodea, and Mantophasmatodea). Conversely, of course, essentially all insect orders primarily have members which are beneficial, in some respects, with the exception of Phthiraptera (lice), Siphonaptera (fleas), and Strepsiptera, the three orders whose members are exclusively parasitic. Insects are considered as pests for a variety of reasons including their:

- direct damage by feeding on crop plants in the field or by infesting stored products
- indirect damage by spreading viral diseases of crop plants (especially by sucking insects such as leafhoppers)
- spreading disease among humans and livestock
- annoyance to humans

Examples

- The Phylloxera plague
- Migratory locust
- Colorado potato beetle
- Boll weevil
- Japanese beetle
- Aphids

- Mosquitoes
- Cockroach
- Western corn rootworm

In the past entomologists working on pest insects attempted to *eradicate* species. This has rarely worked except in islands or controlled environments and raises ethical issues. Over time the language has changed to terms like *control* and *management*. The indiscriminate use of toxic and persistent chemicals and the resurgence of pests in the history of cotton growing in the US has been particularly well studied.

Beneficial Insects

Honey is perhaps the most economically valuable product from insects. Apiculture is a commercial enterprise in most parts of the world and many forest tribes have been dependent on honey as a major source of nutrition. Honeybees can also act as pollinators of crop species. Many predators and parasitoid insects are encouraged and augmented in modern agriculture. Silk is extracted from both reared caterpillars as well as from the wild (producing wild silk). Sericulture deals with the techniques for efficient silkworm rearing and silk production. Although new fabric materials have substituted silk in many applications, it continues to be the material of choice for surgical sutures. Lac was once extracted from scale insects but is now replaced by synthetic substitutes. The dye extracted from cochineal insects was similarly replaced by technological advances. The idea of insects as human food, entomophagy, has been proposed as a solution to meet the growing demand for food, but has not gained widespread acceptance.

HOME STORED PRODUCT ENTOMOLOGY

Home stored product entomology is the study of insects which infest foodstuffs stored in the home. It deals with the prevention, detection and eradication of the pests. The five major culprits considered in this article are flour beetles, the drugstore beetle, the sawtoothed grain beetle, the Indianmeal moth and fruit flies. This is an important branch of forensic entomology because consumers who find products that are contaminated may choose to take legal action against the producers. A suitably qualified entomologist is likely to be able to determine the identity of contaminant species even when no insects are found and the only evidence of infestation is the resulting damage. He or she should also be able to determine whether the foodstuff was contaminated before or after purchase. Companies are required by the U.S. Food and Drug Administration (FDA) to have no more than a certain number of larvae, insects or insect fragments in their product, and it is when this *defect action level* is exceeded that a consumer can prosecute.

Major Stored Product Pests

1. Flour Beetles (Tribolium Castaneum and Tribolium Confusum)

There are two different types of beetles classified as flour beetles: the red flour beetle and the confused flour beetle. Both are similar in physical characteristics. They are flat and oval in shape and usually range around 1/8-inch long. Their exoskeleton is reddish brown with a shiny and smooth texture. The eggs, larvae, and pupa resemble each other closely in physical features as well. The eggs usually tend to be a white color, or at times even colorless. They are very small in size and have a sticky outer covering that causes certain food particles to stick to it. The larvae have six legs, with two pointy projections toward the caudal end. Finally, the pupal stage (a cocoon-like form used to break itself down and emerge as a mature adult) is usually a white or brownish color. The beetle life cycle lasts approximately 3 years or more, with the larval stage ranging anywhere from 20 to over 100 days, and the pupal stage around 8 days. Beetles usually breed in damaged grain, grain dust, high-moisture wheat kernels and flour. The female flour beetle can lay between 300 and 400 eggs during her lifetime [a period of 5 to 8 months]. The flour beetles mainly infest grains, including, but not limited to: cereal, corn meal oats, rice, flour, and crackers. This type of beetle is the most abundant insect pest of flour mills across the United States. Their small size allows them to maneuver through cracks and crevices and get into the home and other areas.

Once they are present in areas with potential food sources, they can infest material such as flour, resulting in a sharp odor or moldy flavor. The red flour beetle is able to fly short distances and the confused flour beetle is unable to fly. While the confused flour beetle is more commonly found in the northern United States, the red flour beetles are more predominant in the southern United States in areas with warmer climates. The red flour beetle is considered the best organism to study genetics. A team composed of Susan Brown and Rob Denell, both biology professors at the University of Kansas, along with Richard Beeman, won financial support to have the beetle's genome sequenced, marking it as one of the earliest sequenced insect genomes, as well as distinguishing the red flour beetle as the only pest insect to be analyzed in this way thus far. The project was funded partially by the US Department of Agriculture, as well as by the National Institutes of Health, and the genome sequencing was led by Stephen Richards at the Human Genome Sequencing Center, Baylor College of Medicine. The red flour beetle's gene map information and genomic sequence are accessible from the National Center for Biotechnical Information. More information is to be published in Nature on 2008-03-27.

2. Drugstore Beetle (Stegobium Paniceum)

This beetle is closely related to the commonly known cigarette beetle. Adult drugstore beetles are cylindrical with lengths ranging from 2.25 to 3.5 mm. These beetles are a reddish brown color and have elytra, sclerotized (hardened) wings that fold back over the abdomen and hinge upwards, allowing the hind wings to come out in order to fly. Females are capable of laying up to 75 eggs during a 13 to 65 days. After the eggs are laid they move into a larval period that can range anywhere from 4 to 20 weeks. After the larval period, drugstore beetle larvae move out of the substrate to build a cocoon and pupate. The pupation period takes a total of 12–18 days. The entire life cycle of the drugstore beetle lasts approximately 2 months but can be as long as 7 months. These stored product pests will infest almost anything readily available. Food products prone to infestation include: flours, dry mixes, breads, cookies and other spices. Non-food material includes: wool, hair, leather and museum specimens. This specific type of beetle has symbiotic yeasts that produce B vitamins, which allow the beetle to survive even when consuming foods of low nutritional value. They are found in areas that have a warmer climate, yet are less plentiful in the tropics than their counterpart, the cigarette beetle.

3. Sawtoothed Grain Beetle (Oryzaephilus Surinamensis)

The sawtoothed grain beetle is closely related to the merchant grain beetle, and is commonly found in kitchen cabinets feeding on items such as cereal, breakfast foods, dried fruits, macaroni, crackers, etc. They are the most common grain and stored product pest in the United States. They are very active and tend to crawl rapidly while searching for food. They are small insects, reaching a length of about 1/8 of an inch. Their name originates from their distinguishable saw-tooth-like projections found on each side of the thorax. The body shape of the beetle is flat, slender, and brown in color. The size and shape of the mandibles allow the beetles to easily break through well sealed and packaged foods. An adult female sawtoothed grain beetle can lay anywhere between 45 to 250 eggs that usually hatch within 3 to 17 days. The larvae look much different from the mature adult. They have a caterpillar-like appearance, with a yellowish coloration to the body and a brown head. The larval period can last as long as 10 weeks but can be as short as 2 weeks. Following the larval instars is the pupal period, which can last one to three weeks. The pupal stage is characterized by the unique process by which these beetles stick together pieces of food material to form a protective covering around their bodies. A fully mature adult beetle, under optimal conditions, can live a maximum of 4 years, a long lifespan for an arthropod.

4. Indianmeal Moth (Plodia Interpunctella)

Indianmeal moths can infest a variety of foods found in the home. Coarsely ground grains, cereals, dried fruits, and herbs are common items the moths have been known to infest. They have also been found in animal food such as dry dog food, fish food and even bird seed. The adult moth is small; generally, their length averages about 3/8 inch, with a 5/8 inch wing span. As adults, the moths are easily identified by an overall grayish, dirty complexion. However, the wing tips have a bronze color that helps differentiate this particular moth from other household moths. The adults have a distinct forewing pattern as well which consists of a light-colored base with about two-thirds of the distal area a red to copper color. The larval stage, or caterpillar, is characterized by a pinkish or yellowish-green body color with a dark brown head. The larval stage of the moth's life cycle is centered on food sources; during the last instar, these larvae are characterized by a movement towards a protected area to pupate. These caterpillars have the capacity to chew through plastic packaging and will often produce silk that loosely binds to food fragments. The pupal stage is generally observed as tiny cocoons that hang from the ceiling; these cocoons can also be found on walls, as well as near the food source. A female can lay over 200 eggs, and will usually die after this process because adults Indian meal moths do not eat.

5. Fruit Flies (Drosophila Melanogaster)

Fruit flies are found near ripened or fermenting fruit.

Tomatoes, melons, squash, grapes and other perishable items brought in from the garden are a common cause of an indoor infestation. Fruit flies can also be attracted to rotting bananas, potatoes, onions and other unrefrigerated produce purchased at the grocery store and taken home. The body of the fruit fly is tan towards the front part of the body and black towards the rear. They usually have red eyes and are about 1/8 inch long. Females have the ability to lay over 500 eggs, usually in fermenting fruit as a food source. The only environment necessary for successful reproduction is a moist film and fermenting material. Generally, fruit flies are a problem during late summer and fall due to their attraction to ripening and fermenting fruits and vegetables. The entire life cycle can be completed in about a week. Unfortunately, because of their ability to fly in and out of the home through windows and screens, they have the capability of contaminating food with bacteria and disease-producing organisms.

Detection of an Infestation

Careful observation is necessary when detecting the culprit of an insect infestation. Each of the five different insects discussed has a unique pattern of destruction. These observations are imperative, as there are not always larvae, pupa, or adults readily available for examination and identification. In the absence of physical specimens, conclusions can be drawn about the probable insect that is infesting the product just by noting the damage done to the particular food. By noting the type of food and the damage done, a nearly accurate conclusion can be drawn about the type of insect causing the damage, allowing a conjecture about the type of control needed. Having an insect specimen and accurately identifying it can lead to eradication, and ultimately, prevention.

Foods commonly infested include:

- Whole or cracked grains (rice)
- Flour, meal, or similar ground grain products

- Spices
- Cereals
- Pasta
- Candy
- Powdered milk
- Nuts (whole or pieces)

Other items include, but aren't limited to: Rodent baits (that contain grain as a feeding attractant), dry pet food, bird seed, grass seed, some powdered soap detergents, dried flowers, potpourri, items stuffed with dried beans or other plant material, and tobacco products. In order to identify an insect, and consequently make a decision about the type of control to be implemented, the type of food must first be noted, especially in the absence of a specimen. Although identifying the food is a general start to begin to identify the insect, it must be remembered that it isn't always the most accurate method. It is mostly used as a guideline, as some insects are more likely than others to be found in certain types of grain, flour, etc. The type of food is not always conclusive to the type of insect found in it, as insects are not extremely picky, and many families and species are found on a wide range of different foodstuffs. Using the infested item as a guideline, noting the type of damage done to the product is the next step. Some insects, like the drugstore beetle, leave telltale tiny holes in the damaged product, while Indianmeal moths are notorious for the spider web-like threads left behind in the food they infest.

These observations can generally lead to a mostly accurate conclusion about the type of insects that are causing the damage, but obviously the most accurate conclusion relies on any specimen found either directly in the stored product or in the vicinity. The larvae, pupae, and adults can be found directly in the product while usually only the pupae and adults are found in the vicinity of the product. It is not practical to assume that any person has knowledge of general entomology, so the following analysis focuses on the five major pests that most commonly infest stored products, beginning with the type of foods infested, signs indicative of a particular insect infestation, and a description of the larvae, pupae, and adults, including behaviour as well as appearance.

Red Flour Beetle Detection

This beetle is similar to the sawtoothed grain beetles in both habits and types of products infested. It is a serious pest in flour mills and wherever cereal products and other dried products are stored and/or processed. Generally, the beetle is attracted to grain with a high moisture content, and usually causes the grain to acquire a grayish tint. The beetle may also impart a bad odor, which then affects the taste of the infested products, as well as encouraging the growth of mold in the grain. This foul odor and taste in the various food products are caused by pheromones and toxic quinone compounds.

Sawtoothed Grain Beetle Detection

The sawtoothed grain beetle feeds on a plethora of feeds, but is not capable of attacking whole or undamaged grains; therefore, it is commonly found in processed grains (flour and meal), dry dog food, dried fruits, candy bars, tobacco, drugs, dried meats, and a variety of other products. It is one of the most common insects found to infest stored products. Larvae are found to develop in flour, dried foods, cereal products, and other stored products.

Drugstore Beetle Detection

These beetles will infest almost anything- they are found most often, however, in flour, bread, spices, breakfast foods, and meal. In the case of an infestation, contaminated products have telltale tunnels which have the appearance of tiny holes. These beetles do not sting, bite, or harm pets or damage a house, yet have the potential, in large infestations, to become a nuisance by flying on doors and windows in heavy populations.

Indianmeal Moth Detection

Indianmeal moths infest both cereal and stored grain products, packaged goods, grain in storage, and surface layers of shelled corn. The most telltale sign of the Indianmeal moth is the silk webbing that the larvae (caterpillars) produce when feeding on the surface of foods. This silk webbing may appear to be or resemble cobwebs inside the products container. Often, a few larvae may be found in the packaging of the product, along with the 'cobwebs', cast skins and frass. Larvae are white worms with black heads, which, when ready to pupate, crawl up the walls of the home in most cases, and are suspended from the ceiling attached by a single silken thread. Most complaints about these moths come during the warmer parts of the year- usually the months of July through August- but the moths have the capability to appear during any month. As with all insects important to stored product entomology, it cannot be automatically assumed that products were previously infested, yet, it is more common for these moths to hitchhike inside a product or in groceries than for the moth to fly into a home through open windows or doors. An important aspect of the Indianmeal moth is that the larvae are the only stage of the insect's life cycle to feed on stored products, the adults do not.

Fruit Fly Detection

Fruit flies are attracted to ripened fruits and vegetables, usually in the kitchen area, but will breed in garbage disposals, empty bottles and cans, wet or damp mops or cleaning rags, and trash containers. The only requirement for these flies to breed is a moist film of fermenting material. Infestations can originate from over-ripened fruits or vegetables that were previously infested, and then brought into the home or from fruit over-ripening in the home. Since adults can also fly from the outside through screened doors or windows it can not always be assumed that the product in question was infested before it was brought into the home. The larvae are found on the inside layer of the fruit, just directly beneath the skin. If the outer layer of the fruit is removed, the rest of the fruit can be salvaged. Fruit flies are primarily a nuisance pest, yet they can infect food with bacteria and other organisms that cause diseases.

FDA Regulations

Defect action levels have been a part of the food industry for nearly a century. The first established defect action level was created in 1911 for mold in tomato pulp. However, limits for insect fragments and larvae were not added until the 1920s on various fruits and vegetables. In 1938, the Federal Food, Drug and Cosmetic Act was established to provide a more defined reference based on strict limitations and methods. Major companies spend a large amount of money every year to aid in the prevention of food contamination. Most of these dollars are well-spent and do, in fact, prevent food from becoming contaminated on a large scale; however, there are many "defects" in consumers' meals on a daily basis. The Food and Drug Administration states, "it is economically impractical to grow, harvest, or process raw products that are totally free of non-hazardous, naturally occurring unavoidable defects". The general public proposes that companies should use more chemicals or pesticides to control this "problem". Though, the amount of pesticide and chemicals necessary to eradicate all insects from foodstuff would pose a threat to any human's health; much more harmful than a controlled quantity of insect and rodent fragments. The Food Defect Action Levels, as proposed by the FDA, is a list of ordinances and guidelines manufacturers and industrial food agencies must abide to ensure the safe service of foodstuff. However, these detection levels are labeled with maximum limitations only. Due to the impossibility of preventing all unavoidable defects in foods, the FDA attempts to prevent these health hazards from reaching a harmful level. Therefore, it is understood and regarded that all manufactures are allowed to have low numbers of insect and rodent hairs present in food, as long as the product is still considered "safe" for human consumption.

Prevention and Eradication

Prevention

To prevent the infestation of foodstuffs by pests of stored products, or "pantry pests", a thorough inspection must be conducted of the food item intended for purchase at the supermarket or the place of purchase. The expiration date of grains and flour must also be noted, as products that sit undisturbed on the shelf for an extended period of time are more likely to become infested. This does not, however, exclude even the freshest of products from being contaminated. Packaging should be inspected for tiny holes that indicate that there might be an infestation. Foodstuffs susceptible to infestation in particular include dried grains, flour, cereals, and whole-grain products. If there is evidence of an insect infestation, the product should not be purchased. The store should be notified immediately, as further infestation must be prevented. Most stores have a plan of action for insect infestations. Bringing an infested product into a pantry or a home leads to a greater degree of infestation. In the home, putting cereal or grain type items in preventative containers will also help to prevent an infestation or the spread of insects from one product to another. Insects can chew through thin plastic, foil, cardboard and other packaging that the product is put in for resale; transferring purchased products into heavy glass containers that can be tightly sealed or heavy plastic containers can improve sanitation and prevent infestation. Using the oldest products first and buying grains and cereals in smaller quantities which can be used up quickly, depending on the size or intake of the family, decreases the chances of infestation. Fruit flies, however, present an entirely different approach to prevention. The primary method to controlling and eliminating fruit flies is to eradicate sources of attraction. Ripened produce should be either eaten, discarded, or refrigerated. Any fruit or vegetable that is damaged or cracked needs to be cut, and the damaged piece discarded in the case that larvae or eggs are present in the area in question. Careful attention must be paid to potential breeding sites that, when forgotten, could cause a massive infestation- all recycling and compost bins must be cleaned, and areas must be checked for forgotten, rotting fruit. Because of their small size, fruit flies are capable of breeding on the inside of the lid of a container. Therefore, it is imperative that when personally canning fruits or vegetables, beer, cider, or wine, that the container is well-sealed. Adults moths can lay eggs under the lid of a jar, allowing the larvae to crawl into the food source when hatched. Homeowners should also outfit their doors and windows with tight mesh screens that prevent the adult fruit flies from flying in from outdoors. Preventative methods and sanitation are the keys to avoiding an infestation or contamination of foodstuffs.

Eradication

Although not seen when groceries are purchased, some products have the possibility of being infested prior to being placed in the pantry. A periodical check of susceptible foodstuffs is necessary, especially in summer months when most insects are more active. In the event that an infestation is discovered, steps must be taken to eradicate the insects. Controlling an infestation is a lengthy process and insects may still be seen, albeit in dwindling numbers, for several weeks. All infested items, as well as noninfested items, must be removed from shelves, thoroughly cleaned and vacuumed. After vacuuming, the waste containing the infested material must be removed and discarded. Items should be checked for beetles, larvae, and pupa; it must be ensured that all food items are inspected as well, and special attention must be paid to items that have been rarely used. The infested items may either be discarded, heated or frozen to kill the insects. If the food is chosen to be discarded, the item must be completely removed from surrounding premises in order to prevent reinfestation.

Freezing products for three to four days or heating them to about 130 to 140 degrees Fahrenheit for 30 to 40 minutes will rid the product of the pests. Decorative ornaments and objects made with plant material and seed located in the vicinity of stored products will increase the risk of reinfestation: insects can feed on those items until they locate stored products.

These items should also be thrown out or eradicated by freezing or heating. Cleaning the area where the infested products were found is advisable, as well. Cleaning with bleach or ammonia, however, will not help with the eradication of the pests. Using a vacuum cleaner to clean the area thoroughly, especially in cracks and corners where insects may hide, will decrease the chances of reinfestation. Because food will be stored in that area again, pesticides are not a good method of eradication. Pesticides can leave a residue that can contaminate food products stored near it. Also, once a pest is inside the container, the pesticides have no effect. If the infestation is so severe that pesticides are the only way to contain the problem, a professional should be contacted immediately. Do not try to apply pesticides to any area where food is stored for human or animal consumption. Contamination can occur and cause illness or more severe conditions. Proper storage and cleanliness are the only ways to prevent an infestation from occurring. Sanitation is the key to prevention and eradication of any pests.

If these insects are found infesting stored products, it is not practical to automatically assume the store or producer is at fault. Although some infestations are not the consumer's fault, producers are held by the FDA action defect levels to ensure their product does not contain more than the allotted amount of insects/insect fragments/larvae. If a stored product is found to be infested by insects, and it is suspected that someone other than the consumer is at fault, a forensic entomologist can be contacted to make a determination. Again, it should not be assumed that these are the only five insects found in a common household pantry, but due to the large number of infestations by these five major groups, it can be safely deduced that it is perhaps one of these five. Stored product entomology is an important forensic field that is important to not only the government and the FDA but the general public, as it is involved in the consumption of food in everyday life.

Other stored product pests:

- Rice weevil *Sitophilus oryzae*
- Maize weevil *Sitophilus zeamais*
- Yellow mealworm *Tenebrio molitor*
- Red Flour Beetle *Tribolium castaneum*
- Confused Flour Beetle *Tribolium confusum*
- Spider beetle *Mezium americanum*
- Fur/Carpet beetle *Attagenus pellio*
- Black carpet beetle *Attagenus unicolor*
- Varied carpet beetle *Anthrenus verbasci*
- Furniture beetle *Anobium punctatum*
- Larder beetle *Dermestes lardarius*
- Khapra beetle *Trogoderma granarium*
- Lesser grain borer *Rhyzopertha dominica*
- Indian meal moth *Plodia interpunctella*
- Meal Moth *Pyralis farinalis*
- Destructive Flour Beetle— *Tribolium destructor*
- Saw-toothed grain beetle *Oryzaephilus surinamensis*
- Merchant grain beetle *Oryzaephilus mercator*
- Australian Spider Beetle *Ptinus tectus*
- Golden Spider Beetle *Niptus hololeucus*
- Red-legged Ham Beetle *Necrobia rufipes*
- Biscuit Beetle *Stegobium paniceum*
- Cigarette Beetle *Lasioderma serricorne*
- Cadelle Beetle *Tenebroides mauritanicus*
- Hide Beetle *Dermestes maculatus*
- Warehouse Beetle *Trogoderma variabile*

- Clothes moths Several species chiefly Webbing Clothing Moth *Tineola bisselliella*; Casemaking Clothes Moth *Tinea pellionella*; *Phereoeca fallax*
- Tapestry Moth *Trichophaga tapetzella*
- Agoumois Moth *Sitotroga cerealella*
- Wine Moth *Oinophila v-flavum*
- Mill Moth *Ephestia kuehniella*
- Warehouse Moth *Ephestia elutella*
- Flour Mite *Acarus siro*
- Silverfish *Lepisma saccharina*
- House cricket *Acheta domesticus*

REFERENCES

Anderson G.S., VanLaerhoven S.L. (1996). "Initial studies on insect succession on carrion in southwestern British Columbia". *Journal of Forensic Sciences* 41 (4): 617–625.

Benecke M. (2001). "A brief history of forensic entomology". *Forensic. Sci. Int.* 120 (1–2): 2–14.

Carloye, Lisa (2003). "Of Maggots & Murder: Forensic Entomology in the Classroom". *The American Biology Teacher* 65 (5): 360.

Catts E.P., Goff M.L. (1992). "Forensic entomology in criminal investigations". *Annual Review of Entomology* 37: 253–272.

D.S. Hill (1983). *Agricultural Insect Pests of the Tropics and their Control*. Cambridge University Press.

Eberhardt TL, Elliot D.A. (2008). "A preliminary investigation of insect colonisation and succession on remains in New Zealand". *Forensic Sci. Int.* 176 (2–3): 217–223.

Karl von Frisch (December 12, 1973). *Decoding the Language of the Bee*, Nobel Lecture.

Lewis, Donald (1995-08). "Insect Pests of Stored Products". *Iowa State University*. University Cooperative Extension.

Liddell, Henry George and Robert Scott (1980). *A Greek-English Lexicon (Abridged Edition)*. United Kingdom: Oxford University Press.

Marlowe, Stephen (1998). "Introduction." The Fall of the House of Usher and Other Tales. New York: New American Library.

Metcalf, C.L. and W.P. Flint (1925). Destructive and Useful Insects.

O'Donnell, Sean (January 1998). "Reproductive Caste Determination In Eusocial Wasps (Hymenoptera: Vespidae)". *Annual Review of Entomology* 43: 323–346.

Payne, J.A. (1965). "A summer carrion study of the baby pig sus scrofa Linnaeus". *Ecology* 46 (5): 511–523.

Poe, Edgar Allan (1984). The Complete Works of Edgar Allan Poe. New York: Doubleday, p. 815

Potter, Michael F. (1994-01). "Fruit Flies". *University of Kentucky College of Agriculture*. University of Kentucky College of Agriculture.

Rodriguez W.C., Bass W.M. (1983). "Insect activity and its relationship to decay rates of human cadavers in East Tennessee". *Journal of Forensic Sciences* 28 (2): 423–432.

Schoenly K., Reid, W. (1987). "Dynamics of heterotrophic succession in carrion arthropod assemblages: discrete seres or a continuum of change?". *Oecologia* 73 (2): 192–202.

Tarone, Aaron M.; Jennings, Kimberley C.; Foran, David R. (November 2007). "Aging Blow Fly Eggs Using Gene Expression: A Feasibility Study". *Journal of Forensic Sciences* 52 (6): 1350–1354.

Tsutsui, Neil D.; Suarez Andrew V. (2003). "The Colony Structure and Population Biology of Invasive Ants". *Conservation Biology* 17 (1): 48–58.

Wells, D. and Sperling Felix A.H. (3 March 2008). "DNA-based identification of forensically important Chrysomyinae (Diptera: Calliphoridae)"Forensic Science International Volume 120, Issues 1-215 August 2001 110-115.

2

Introduction to Insects, Insect Ecology and Other Related Areas

INSECT

Insects are a class of living creatures within the arthropods that have a chitinous exoskeleton, a three-part body (head, thorax, and abdomen), three pairs of jointed legs, compound eyes, and two antennae. They are among the most diverse groups of animals on the planet and include more than a million described species and represent more than half of all known living organisms. The number of extant species is estimated at between six and ten million, and potentially represent over 90% of the differing metazoan life forms on Earth. Insects may be found in nearly all environments, although only a small number of species occur in the oceans, a habitat dominated by another arthropod group, the crustaceans. The life cycles of insects vary but most hatch from eggs. Insect growth is constrained by the inelastic exoskeleton and development involves a series of molts. The immature stages can differ from the adults in structure, habit and habitat and can include a passive pupal stage in those groups that undergo complete metamorphosis. Insects that undergo incomplete metamorphosis lack a pupal stage and adults develop through a series of nymphal stages. The higher level relationship of the hexapoda is unclear. Fossilized insects of enormous size have been found from the Paleozoic Era, including giant dragonflies with wingspans of 55 to 70 cm

(22–28 in). The most diverse insect groups appear to have coevolved with flowering plants.

Insects typically move about by walking, flying or occasionally swimming. As it allows for rapid yet stable movement, many insects adopt a tripedal gait in which they walk with their legs touching the ground in alternating triangles. Insects are the only invertebrates to have evolved flight. Many insects spend at least part of their life underwater, with larval adaptations that include gills and some adult insects are aquatic and have adaptations for swimming. Some species, like water striders, are capable of walking on the surface of water. Insects are mostly solitary, but some insects, such as certain bees, ants, and termites are social and live in large, well-organized colonies. Some insects, like earwigs, show maternal care, guarding their eggs and young. Insects can communicate with each other in a variety of ways. Male moths can sense the pheromones of female moths over distances of many kilometers. Other species communicate with sounds: crickets stridulate, or rub their wings together, to attract a mate and repel other males. Lampyridae in the beetle order Coleoptera communicate with light. Humans regard certain insects as pests and attempt to control them using insecticides and a host of other techniques. Some insects damage crops by feeding on sap, leaves or fruits, a few bite humans and livestock, alive and dead, to feed on blood and some are capable of transmitting diseases to humans, pets and livestock. Many other insects are considered ecologically beneficial and a few provide direct economic benefit. Silkworms and bees have been used extensively by humans for the production of silk and honey, respectively.

External Morphology

General Body Plan

Insects have segmented bodies supported by an exoskeleton, a hard outer covering made mostly of chitin. The segments of the body are organized into three distinctive but interconnected units, or tagmata: a head, a thorax, and an abdomen. The head supports a pair of sensory antennae, a pair of compound eyes, and, if present, one to three simple eyes (or ocelli) and three sets of variously modified appendages that form the mouthparts. The thorax has six segmented legs—one pair each for the prothorax, mesothorax and the metathorax segments making up the thorax—and, if present in the species, two or four wings. The abdomen consists of eleven segments, though in a few species of insects these segments may be fused together or reduced in size. The abdomen also contains most of the digestive, respiratory, excretory and reproductive internal structures. There is considerable variation and many adaptations in the body parts of insects especially wings, legs, antenna, mouth-parts etc.

Exoskeleton

Insect outer skeleton, the cuticle, is made up of two layers: the epicuticle, which is a thin and waxy water resistant outer layer and contains no chitin, and a lower layer called the procuticle. The procuticle is chitinous and much thicker than the epicuticle and has two layers: an outer layer known as the exocuticle and an inner layer known as the endocuticle. The tough and flexible endocuticle is built from numerous layers of fibrous chitin and proteins, criss-crossing each others in a sandwich pattern, while the exocuticle is rigid and hardened. The exocuticle is greatly reduced in many soft-bodied insects (e.g., caterpillars), especially during their larval stages. Insects are the only invertebrates to have developed active flight capability, and this has played an important role in their success. Their muscles are able to contract multiple times for each single nerve impulse, allowing the wings to beat faster than would ordinarily be possible. Having their muscles attached to their exoskeletons is more efficient and allows more muscle connections; crustaceans also use the same method, though all spiders use hydraulic pressure to extend their legs, a system inherited from their pre-arthropod ancestors. Unlike insects, though, most aquatic crustaceans are biomineralized with calcium carbonate extracted from the water.

Internal Morphology

Nervous System

The nervous system of an insect can be divided into a brain and a ventral nerve cord. The head capsule is made up of six fused segments, each with a pair of ganglia, or a cluster of nerve cells outside of the brain. The first three pairs of ganglia are fused into the brain, while the three following pairs are fused into a structure of three pairs of ganglia under the insect's esophagus, called the subesophageal ganglion. The thoracic segments have one ganglion on each side, which are connected into a pair, one pair per segment. This arrangement is also seen in the abdomen but only in the first eight segments. Many species of insects have reduced numbers of ganglia due to fusion or reduction. Some cockroaches have just six ganglia in the abdomen, whereas the wasp *Vespa crabro* has only two in the thorax and three in the abdomen. Some insects, like the house fly *Musca domestica*, have all the body ganglia fused into a single large thoracic ganglion. At least a few insects have nociceptors, cells that detect and transmit sensations of pain. This was discovered in 2003 by studying the variation in reactions of larvae of the common fruitfly Drosophila to the touch of a heated probe and an unheated one. The larvae reacted to the touch of the heated probe with a stereotypical rolling behaviour that was not exhibited when the larvae were touched by the unheated probe. Although nociception has been demonstrated in insects, there is not a consensus that insects feel pain consciously.

Digestive System

An insect uses its digestive system to extract nutrients and other substances from the food it consumes. Most of this food is ingested in the form of macromolecules and other complex substances like proteins, polysaccharides, fats, and nucleic acids. These macromolecules must be broken down by catabolic reactions into smaller molecules like amino acids and simple sugars before being used by cells of the body for energy, growth, or reproduction. This break-down process is known as digestion. The main structure of an insect's digestive system is a long enclosed tube called the alimentary canal, which runs lengthwise through the body. The alimentary canal directs food unidirectionally from the mouth to the anus. It has three sections, each of which performs a different process of digestion. In addition to the alimentary canal, insects also have paired salivary glands and salivary reservoirs. These structures usually reside in the thorax, adjacent to the foregut. The salivary glands (element 30 in numbered diagram) in an insect's mouth produce saliva. The salivary ducts lead from the glands to the reservoirs and then forward through the head to an opening called the salivarium, located behind the hypopharynx. By moving its mouthparts (element 32 in numbered diagram) the insect can mix its food with saliva. The mixture of saliva and food then travels through the salivary tubes into the mouth, where it begins to break down. Some insects, like flies, have extra-oral digestion. Insects using extra-oral digestion expel digestive enzymes onto their food to break it down. This strategy allows insects to extract a significant proportion of the available nutrients from the food source. The gut is where almost all of insects' digestion takes place. It can be divided into the foregut, midgut and hindgut.

Foregut

The first section of the alimentary canal is the foregut (element 27 in numbered diagram), or stomodaeum. The foregut is lined with a cuticular lining made of chitin and proteins as protection from tough food. The foregut includes the buccal cavity (mouth), pharynx, esophagus, and Crop and proventriculus (any part may be highly modified) which both store food and signify when to continue passing onward to the midgut. Here, digestion starts as partially chewed food is broken down by saliva from the salivary glands. As the salivary glands produce fluid and carbohydrate-digesting enzymes (mostly amylases), strong muscles in the pharynx pump fluid into the buccal cavity, lubricating the food like the salivarium does, and helping blood feeders, and xylem and phloem feeders. From there, the pharynx passes food to the esophagus, which could be just a simple tube passing it on to the crop and proventriculus, and then onward to the midgut, as in most insects. Alternately, the foregut may expand into a very enlarged crop and proventriculus, or the crop could just be a diverticulum, or fluid filled structure, as in some Diptera species.

Midgut

Once food leaves the crop, it passes to the midgut (element 13 in numbered diagram), also known as the mesenteron, where the majority of digestion takes place. Microscopic projections from the midgut wall, called microvilli, increase the surface area of the wall and allow more nutrients to be absorbed; they tend to be close to the origin of the midgut. In some insects, the role of the microvilli and where they are located may vary. For example, specialized microvilli producing digestive enzymes may more likely be near the end of the midgut, and absorption near the origin or beginning of the midgut.

Hindgut

In the hindgut (element 16 in numbered diagram), or proctodaeum, undigested food particles are joined by uric acid to form fecal pellets. The rectum absorbs 90% of the water in these fecal pellets, and the dry pellet is then eliminated through the anus (element 17), completing the process of digestion. The uric acid is formed using hemolymph waste products diffused from the Malpighian tubules (element 20). It is then emptied directly into the alimentary canal, at the junction between the midgut and hindgut. The number of Malpighian tubules possessed by a given insect varies between species, ranging from only two tubules in some insects to over 100 tubules in others.

Respiration and Circulation

Insect respiration is accomplished without lungs. Instead, the insect respiratory system uses a system of internal tubes and sacs through which gases either diffuse or are actively pumped, delivering oxygen directly to tissues that need it via their trachea (element 8 in numbered diagram). Since oxygen is delivered directly, the circulatory system is not used to carry oxygen, and is therefore greatly reduced. The insect circulatory system has no veins or arteries, and instead consists of little more than a single, perforated dorsal tube which pulses peristaltically. Toward the thorax, the dorsal tube (element 14) divides into chambers and acts like the insect's heart. The opposite end of the dorsal tube is like the aorta of the insect circulating the hemolymph, arthropods' fluid analog of blood, inside the body cavity. Air is taken in through openings on the sides of the abdomen called spiracles. There are many different patterns of gas exchange demonstrated by different groups of insects. Gas exchange patterns in insects can range from continuous and diffusive ventilation, to discontinuous gas exchange. During continuous gas exchange, oxygen is taken in and carbon dioxide is released in a continuous cycle. In discontinuous gas exchange, however, the insect takes in oxygen while it is active and small amounts of carbon dioxide are released when the insect is at rest. Diffusive ventilation is simply a form of continuous gas exchange that occurs by diffusion rather than physically taking in the oxygen. Some species of insect that are submerged also have adaptations to aid in respiration. As larvae, many insects have gills that can extract oxygen dissolved in water, while others need to rise to the water surface to replenish air supplies which may be held or trapped in special structures.

Reproduction and Development

The majority of insects hatch from eggs. The fertilization and development takes place inside the egg, enclosed by a shell (chorion). Some species of insects, like the cockroach

Blaptica dubia, as well as juvenile aphids and tsetse flies, are ovoviviparous. The eggs of ovoviviparous animals develop entirely inside the female, and then hatch immediately upon being laid. Some other species, such as those in the genus of cockroaches known as *Diploptera*, are viviparous, and thus gestate inside the mother and are born alive. Some insects, like parasitic wasps, show polyembryony, where a single fertilized egg divides into many and in some cases thousands of separate embryos. Other developmental and reproductive variations include haplodiploidy, polymorphism, paedomorphosis or peramorphosis, sexual dimorphism, parthenogenesis and more rarely hermaphroditism. In haplodiploidy, which is a type of sex-determination system, the offspring's sex is determined by the number of sets of chromosomes an individual receives. This system is typical in bees and wasps. Polymorphism is the where a species may have different *morphs* or *forms*, as in the oblong winged katydid, which has four different varieties: green, pink, and yellow or tan. Some insects may retain phenotypes that are normally only seen in juveniles; this is called paedomorphosis.

In peramorphosis, an opposite sort of phenomenon, insects take on previously unseen traits after they have matured into adults. Many insects display sexual dimorphism, in which males and females have notably different appearances, such as the moth *Orgyia recens* as an exemplar of sexual dimorphism in insects. Some insects use parthenogenesis, a process in which the female can reproduce and give birth without having the eggs fertilized by a male. Many aphids undergo a form of parthenogenesis, called cyclical parthenogenesis, in which they alternate between one or many generations of asexual and sexual reproduction. In summer, aphids are generally female and parthenogenetic; in the autumn, males may be produced for sexual reproduction. Other insects produced by parthenogenesis are bees, wasps, and ants, in which they spawn males. However, overall, most individuals are female, which are produced by fertilization. The males are haploid and the females are diploid. More rarely, some insects display hermaphroditism, in which a given individual has both male and female reproductive organs. Insect life-histories show adaptations to withstand cold and dry conditions. Some temperate region insects are capable of activity during winter, while some others migrate to a warmer climate or go into a state of torpor. Still other insects have evolved mechanisms of diapause that allow eggs or pupae to survive these conditions.

Metamorphosis

Metamorphosis in insects is the biological process of development all insects must undergo. There are two forms of metamorphosis: incomplete metamorphosis and complete metamorphosis.

Incomplete Metamorphosis

Insects that show hemimetabolism, or incomplete metamorphosis, change gradually by undergoing a series of molts. An insect molts when it outgrows its exoskeleton, which does not stretch and would otherwise restrict the insect's growth. The molting process begins as the insect's epidermis secretes a new epicuticle. After this new epicuticle is secreted, the epidermis releases a mixture of enzymes that digests the endocuticle and thus detaches the old cuticle. When this stage is complete, the insect makes its body swell by taking in a large quantity of water or air, which makes the old cuticle split along predefined weaknesses where the old exocuticle was thinnest. Other arthropods have a much different process and only molt; though must accommodate for the difference in exoskeleton structure and make up with other enzymes. Immature insects that go through incomplete metamorphosis are called nymphs or in the case of dragonflies and damselflies as naiads. Nymphs are similar in form to the adult except for the presence of wings, which are not developed until adulthood. With each molt, nymphs grow larger and become more similar in appearance to adult insects.

Complete Metamorphosis

Holometabolism, or complete metamorphosis, is where the insect changes all in four stages, an egg or embryo, a larva, a pupa, and the adult or imago. In these species, egg hatches to produce a larva, which is generally worm-like in form. This worm-like form can be one of several varieties: eruciform (caterpillar-like), scarabaeiform (grub-like), campodeiform (elongated, flattened, and active), elateriform (wireworm-like) or vermiform (maggot-like). The larva grows and eventually becomes a pupa, a stage marked by reduced movement and often sealed within a cocoon. There are three types of pupae: obtect, exarate or coarctate. Obtect pupae are compact, with the legs and other appendages enclosed. Exarate pupae have their legs and other appendages free and extended. Coarctate pupae develop inside the larval skin. Insects undergo considerable change in form during the pupal stage, and emerge as adults. Butterflies are a well known example of an insects that undergo complete metamorphosis, although most insects use this life cycle. Some insects have evolved this system to hypermetamorphosis. Some of the oldest and most successful insect groups, such Endopterygota, use a system of complete metamorphosis. Strangely though, complete metamorphosis is unique to certain insect orders, like Diptera, Lepidoptera, and Hymenoptera, and no other arthropods undergo it, but incomplete metamorphosis.

Senses and Communication

Many insects possess very sensitive and/or specialized organs of perception. Some insects such as bees can perceive ultraviolet wavelengths, or detect polarized light, while the antennae of male moths can detect the pheromones of female moths over distances of many kilometers. There is a pronounced tendency for there to be a trade-off between visual acuity and chemical or tactile acuity, such that most insects with well-developed eyes have reduced or simple antennae, and vice-versa. There are a variety of different mechanisms by which insects perceive sound, while the patterns are not universal, insects can generally hear sound if they can produce it. Different insect species can have varying hearing, though most insects can hear only a narrow range of frequencies related to the frequency of the sounds they can produce. Mosquitoes have been found to hear up to 2 MHz., and some grasshoppers can hear up to 50 MHz. Certain predatory and parasitic insects can detect the characteristic sounds made by their prey or hosts, respectively. For instance, some nocturnal moths can perceive the ultrasonic emissions of bats, which helps them avoid predation. Insects that feed on blood have special sensory structures that can detect infrared emissions, and use them to home in on their hosts. Some insects display a rudimentary sense of numbers, such as the solitary wasps that prey upon a single species. The mother wasp lays her eggs in individual cells and provides each egg with a number of live caterpillars on which the young feed when hatched. Some species of wasp always provide five, others twelve, and others as high as twenty-four caterpillars per cell. The number of caterpillars is different among species, but always the same for each sex of larva. The male solitary wasp in the genus *Eumenes* is smaller than the female, so the mother of one species supplies him with only five caterpillars; the larger female receives ten caterpillars in her cell.

Light Production and Vision

A few insects, such as members of the families Poduridae and Onychiuridae (Collembola), Mycetophilidae (Diptera), and the beetle families Lampyridae, Phengodidae, Elateridae and Staphylinidae are bioluminescent. The most familiar group are the fireflies, beetles of the family Lampyridae. Some species are able to control this light generation to produce flashes. The function varies with some species using them to attract mates, while others use them to lure prey. Cave dwelling larvae of *Arachnocampa* (Mycetophilidae, Fungus gnats) glow to lure small flying insects into sticky strands of silk. Some fireflies of the genus *Photuris* mimic the flashing of female *Photinus* species to attract males of that species, which are then captured and devoured. The colors of emitted light vary from dull blue (*Orfelia fultoni*, Mycetophilidae) to the familiar greens and the rare reds (*Phrixothrix tiemanni*, Phengodidae). Most insects, except some species of cave dwelling crickets, are able to perceive light and dark. Many species have acute vision capable of detecting minute movements. The eyes include simple eyes or ocelli as well as compound eyes of varying sizes. Many species are able to detect light in the infrared, ultraviolet and the visible light wavelengths. Color vision has been demonstrated in many species and phylogenetic analysis suggests that UV-green-blue trichromacy existed from at least the Devonian period between 416 and 359 million years ago.

Sound Production and Hearing

Insects were the earliest organisms to produce and sense sounds. Insects make sounds mostly by mechanical action of appendages. In grasshoppers and crickets, this is achieved by stridulation. Cicadas make the loudest sounds among the insects by producing and amplifying sounds with special modifications to their body and musculature. The African cicada *Brevisana brevis* has been measured at 106.7 decibels at a distance of 50 cm (20 in). Some insects, such as the hawk moths and Hedylid butterflies, can hear ultrasound and take evasive action when they sense that they have been detected by bats. Some moths produce ultrasonic clicks that were once thought to have a role in jamming bat echolocation. The ultrasonic clicks were subsequently found to be produced mostly by unpalatable moths to warn bats, just as warning colorations are used against predators that hunt by sight. Some otherwise palatable moths have evolved to mimic these calls. Very low sounds are also produced in various species of Coleoptera, Hymenoptera, Lepidoptera, Mantodea, and Neuroptera. These low sounds are simply the sounds made by the insect's movement. Through microscopic stridulatory structures located on the insect's muscles and joints, the normal sounds of the insect moving are amplified and can be used to warn or communicate with other insects. Most sound-making insects also have tympanal organs that can perceive airborne sounds. Some species in Hemiptera, such as the corixids (water boatmen), are known to communicate via underwater sounds.

Most insects are also able to sense vibrations transmitted through surfaces. For example, an insect is caught in a spider web and struggles to escape. The vibrations it produces are sensed by the spider, who is alerted to its presence. Through these vibrations, the spider can tell where on the web the insect is located, as well as how big it is. Communication using surface-borne vibrational signals is more widespread among insects because of size constraints in producing air-borne sounds. Insects cannot effectively produce low-frequency sounds, and high-frequency sounds tend to disperse more in a dense environment (such as foliage), so insects living in such environments communicate primarily using substrate-borne vibrations. The mechanisms of production of vibrational signals are just as diverse as those for producing sound in insects. Some species use vibrations for communicating within members of the same species, such as to attract mates as in the songs of the shield bug *Nezara viridula*. Vibrations can also be used to communicate between entirely different species; lycaenid (gossamer-winged butterfly) caterpillars which are myrmecophilous (living in a mutualistic association with ants) communicate with ants in this way. The Madagascar hissing cockroach has the ability to press air through its spiracles to make a hissing noise as a sign of aggression; the Death's-head Hawkmoth makes a squeaking noise by forcing air out of their pharynx when agitated, which may also reduce aggressive worker honey bee behaviour when the two are in close proximity.

Chemical Communication

In addition to the use of sound for communication, a

wide range of insects have evolved chemical means for communication. These chemicals, termed semiochemicals, are often derived from plant metabolites include those meant to attract, repel and provide other kinds of information. Pheromones, a type of semiochemical, are used for attracting mates of the opposite sex, for aggregating conspecific individuals of both sexes, for deterring other individuals from approaching, to mark a trail, and to trigger aggression in nearby individuals. Allomonea benefit their producer by the effect they have upon the receiver. Kairomones benefit their receiver instead of their producer. Synomones benefit the producer and the receiver. While some chemicals are targeted at individuals of the same species, others are used for communication across species. The use of scents is especially well known to have developed in social insects.

Social Behaviour

Social insects, such as termites, ants and many bees and wasps, are the most familiar species of eusocial animal. They live together in large well-organized colonies that may be so tightly integrated and genetically similar that the colonies of some species are sometimes considered superorganisms. It is sometimes argued that the various species of honey bee are the only invertebrates (and indeed one of the few non-human groups) to have evolved a system of abstract symbolic communication where a behaviour is used to *represent* and convey specific information about something in the environment. In this communication system, called dance language, the angle at which a bee dances represents a direction relative to the sun, and the length of the dance represents the distance to be flown. Only insects which live in nests or colonies demonstrate any true capacity for fine-scale spatial orientation or homing. This can allow an insect to return unerringly to a single hole a few millimeters in diameter among thousands of apparently identical holes clustered together, after a trip of up to several kilometers' distance. In a phenomenon known as philopatry, insects that hibernate have shown the ability to recall a specific location up to a year after last viewing the area of interest. A few insects seasonally migrate large distances between different geographic regions (e.g., the overwintering areas of the Monarch butterfly).

Care of Young

Most insects lead short lives as adults, and rarely interact with one another except to mate or compete for mates. A small number exhibit some form of parental care, where they will at least guard their eggs, and sometimes continue guarding their offspring until adulthood, and possibly even feeding them. Another simple form of parental care is to construct a nest (a burrow or an actual construction, either of which may be simple or complex), store provisions in it, and lay an egg upon those provisions. The adult does not contact the growing offspring, but it nonetheless does provide food. This sort of care is typical of bees and various types of wasps.

Locomotion

Flight

Insects are the only group of invertebrates to have developed flight. The evolution of insect wings has been a subject of debate. Some entomologists suggest that the wings are from paranotal lobes, or extensions from the insect's exoskeleton called the nota, called the *paranotal theory*. Other theories are based on a pleural origin. The pleuron is membrane on the sides of the thorax. These theories include suggestions that wings originated from modified gills, spiracular flaps or as from an appendage of the epicoxa. The *epicoxal theory* suggests the insect wings are modified epicoxal exites, a modified appendage at the base of the legs or coxa. In the Carboniferous age, some of the *Meganeura* dragonflies had as much as a 50 cm (20 in) wide wingspan. The appearance of gigantic insects has been found to be consistent with high atmospheric oxygen. The respiratory system of insects constrains their size, however the high oxygen in the atmosphere allowed larger sizes. The largest flying insects today are much smaller and include several moth species such as the Atlas moth and the White Witch (*Thysania agrippina*). Insect flight has been a topic of great interest in aerodynamics due partly to the inability of steady-state theories to explain the lift generated by the tiny wings of insects. Unlike birds, many small insects are swept along by the prevailing winds although many of the larger insects are known to make migrations. Aphids are known to be transported long distances by low-level jet streams. As such, fine line patterns associated with converging winds within weather radar imagery, like the WSR-88D radar network, often represent large groups of insects.

Walking

Many adult insects use six legs for walking and have adopted a tripedal gait. The tripedal gait allows for rapid walking while always having a stable stance and has been studied extensively in cockroaches. The legs are used in alternate triangles touching the ground. For the first step, the middle right leg and the front and rear left legs are in contact with the ground and move the insect forward, while the front and rear right leg and the middle left leg are lifted and moved forward to a new position. When they touch the ground to form a new stable triangle the other legs can be lifted and brought forward in turn and so on. The purest form of the tripedal gait is seen in insects moving at high speeds. However, this type of locomotion is not rigid and insects can adapt a variety of gaits. For example, when moving slowly, turning, or avoiding obstacles, four or more feet may be touching the ground. Insects can also adapt their gait to cope with the loss of one or more limbs. Cockroaches are among the fastest insect runners and, at full speed, adopt a bipedal run to reach a high velocity in proportion to their body size. As cockroaches move very quickly, they need to be video recorded at several hundred frames per second to reveal their gait. More sedate locomotion is seen in the stick insects or walking sticks (Phasmatodea). A few insects have evolved to walk on the surface of the water, especially the bugs of the Gerridae family, commonly known as water striders. A few species of ocean-skaters in the genus *Halobates* even live on the surface of open oceans, a habitat that has few insect species.

Use in Robotics

Insect walking is of particular interest as an alternative form of locomotion in robots. The study of insects and bipeds has a significant impact on possible robotic methods of transport. This may allow new robots to be designed that can traverse terrain that robots with wheels may be unable to handle.

Swimming

A large number of insects live either parts or the whole of their lives underwater. In many of the more primitive orders of insect, the immature stages are spent in an aquatic environment. Some groups of insects, like certain water beetles, have aquatic adults as well. Many of these species have adaptations to help in under-water locomotion. Water beetles and water bugs have legs adapted into paddle-like structures. Dragonfly naiads use jet propulsion, forcibly expelling water out of their rectal chamber. Some species like the water striders are capable of walking on the surface of water. They can do this because their claws are not at the tips of the legs as in most insects, but recessed in a special groove further up the leg; this prevents the claws from piercing the water's surface film. Other insects such as the Rove beetle *Stenus* are known to emit pygidial gland secretions that reduce surface tension making it possible for them to move on the surface of water by Marangoni propulsion (also known by the German term *Entspannungsschwimmen*).

Phylogeny and Systematics

The evolutionary relationships of insects to other animal groups remain unclear. Although more traditionally grouped with millipedes and centipedes, evidence has emerged favouring closer evolutionary ties with crustaceans. In the Pancrustacea theory, insects, together with Remipedia and Malacostraca, make up a natural clade. Other terrestrial arthropods, such as centipedes, millipedes, scorpions and spiders, are sometimes confused with insects since their body plans can appear similar, sharing (as do all arthropods) a jointed exoskeleton. However, upon closer examination their features differ significantly; most noticeably they do not have the six legs characteristic of adult insects. The higher-level phylogeny of the arthropods continues to be a matter of debate and research. In 2008, researchers at Tufts University uncovered what they believe is the world's oldest known full-body impression of a primitive flying insect, a 300 million-year-old specimen from the Carboniferous Period. The oldest definitive insect fossil is the Devonian *Rhyniognatha hirsti*, from the 396 million year old Rhynie chert. It may have superficially resembled a modern-day silverfish insect. This species already possessed dicondylic mandibles (two articulations in the mandible), a feature associated with winged insects, suggesting that wings may already have evolved at this time. Thus, the first insects probably appeared earlier, in the Silurian period.

There have been four super radiations of insects: beetles (evolved ~300 million years ago), flies (evolved ~250 million years ago), moths and wasps (evolved ~150 million years ago). These four groups account for the majority of described species. The flies and moths along with the fleas evolved from the Mecoptera. The origins of insect flight remain obscure, since the earliest winged insects currently known appear to have been capable fliers. Some extinct insects had an additional pair of winglets attaching to the first segment of the thorax, for a total of three pairs. As of 2009, there is no evidence that suggests that the insects were a particularly successful group of animals before they evolved to have wings. Late Carboniferous and Early Permian insect orders include both extant groups and a number of

Paleozoic species, now extinct. During this era, some giant dragonfly-like forms reached wingspans of 55 to 70 cm (22 to 28 in) making them far larger than any living insect. This gigantism may have been due to higher atmospheric oxygen levels that allowed increased respiratory efficiency relative to today. The lack of flying vertebrates could have been another factor. Most extinct orders of insects developed during the Permian period that began around 270 million years ago. Many of the early groups became extinct during the Permian-Triassic extinction event, the largest mass extinction in the history of the Earth, around 252 million years ago.

The remarkably successful Hymenopterans appeared as long as 146 million years ago in the Cretaceous period, but achieved their wide diversity more recently in the Cenozoic era, which began 66 million years ago. A number of highly successful insect groups evolved in conjunction with flowering plants, a powerful illustration of coevolution. Many modern insect genera developed during the Cenozoic. Insects from this period on are often found preserved in amber, often in perfect condition. The body plan, or *morphology*, of such specimens is thus easily compared with modern species. The study of fossilized insects is called paleoentomology.

Evolutionary Relationships

Insects are prey for a variety of organisms, including terrestrial vertebrates. The earliest vertebrates on land existed 400 million years ago and were large amphibious piscivores, through gradual evolutionary change, insectivory was the next diet type to evolve. Insects were among the earliest terrestrial herbivores and acted as major selection agents on plants. Plants evolved chemical defenses against this herbivory and the insects in turn evolved mechanisms to deal with plant toxins. Many insects make use of these toxins to protect themselves from their predators. Such insects often advertise their toxicity using warning colors. This successful evolutionary pattern has also been utilized by mimics. Over time, this has led to complex groups of coevolved species. Conversely, some interactions between plants and insects, like pollination, are beneficial to both organisms. Coevolution has led to the development of very specific mutualisms in such systems.

Taxonomy

Traditional morphology-based or appearance-based systematics has usually given Hexapoda the rank of superclass, and identified four groups within it: insects (Ectognatha), springtails (Collembola), Protura and Diplura, the latter three being grouped together as Entognatha on the basis of internalized mouth parts. Supraordinal relationships have undergone numerous changes with the advent of methods based on evolutionary history and genetic data. A recent theory is that Hexapoda is polyphyletic (where the last common ancestor was not a member of the group), with the entognath classes having separate evolutionary histories from Insecta. Many of the traditional appearance-based taxa have been shown to be paraphyletic, so rather than using ranks like subclass, superorder and infraorder, it has proved better to use monophyletic groupings (in which the last common ancestor is a member of the group). The following represents the best supported monophyletic groupings for the Insecta. Insects can be divided into two groups historically treated as subclasses: wingless insects, known as Apterygota, and winged insects, known as Pterygota. The Apterygota consist of the primitively wingless order of the silverfish (Thysanura). Archaeognatha make up the Monocondylia based on the shape of their mandibles, while Thysanura and Pterygota are grouped together as Dicondylia.

It is possible that the Thysanura themselves are not monophyletic, with the family Lepidotrichidae being a sister group to the Dicondylia (Pterygota and the remaining Thysanura). Paleoptera and Neoptera are the winged orders of insects differentiated by the presence of hardened body parts called sclerites; also, in Neoptera, muscles that allow their wings to fold flatly over the abdomen. Neoptera can further be divided into incomplete metamorphosis-based (Polyneoptera and Paraneoptera) and complete metamorphosis-based groups. It has proved difficult to clarify the relationships between the orders in Polyneoptera because of constant new findings calling for revision of the taxa. For example, Paraneoptera has turned out to be more closely related to Endopterygota than to the rest of the Exopterygota. The recent molecular finding that the traditional louse orders Mallophaga and Anoplura are derived from within Psocoptera has led to the new taxon Psocodea. Phasmatodea and Embiidina have been suggested to form Eukinolabia. Mantodea, Blattodea and Isoptera are thought to form a monophyletic group termed Dictyoptera.

It is likely that Exopterygota is paraphyletic in regard to Endopterygota. Matters that have had a lot of controversy include Strepsiptera and Diptera grouped together as Halteria based on a reduction of one of the wing pairs – a position not well-supported in the entomological community. The Neuropterida are often lumped or split on the whims of the taxonomist. Fleas are now thought to be closely related to boreid mecopterans. Many questions remain to be answered when it comes to basal relationships amongst endopterygote orders, particularly Hymenoptera. The study of the classification or taxonomy of any insect is called systematic entomology. If one works with a more specific order or even a family, the term may also be made specific to that order or family, for example systematic dipterology.

Relationship to Humans

Many insects are considered pests by humans. Insects commonly regarded as pests include those that are parasitic (mosquitoes, lice, bed bugs), transmit diseases (mosquitoes, flies), damage structures (termites), or destroy agricultural goods (locusts, weevils). Many entomologists are involved in various forms of pest control, as in research for companies to produce insecticides, but increasingly relying on methods of biological pest control, or biocontrol. Biocontrol uses one organism to reduce the population density of another organism — the pest — and is considered a key element of integrated pest management. Despite the large amount of effort focused at controlling insects, human attempts to kill pests with insecticides can backfire. If used carelessly the poison can kill all kinds of organisms in the area, including insects' natural predators such as birds, mice, and other insectivores. The effects of DDT's use exemplifies how some insecticides can threaten wildlife beyond intended populations of pest insects.

Although pest insects attract the most attention, many insects are beneficial to the environment and to humans. Some insects, like wasps, bees, butterflies, and ants, pollinate flowering plants. Pollination is a mutualistic relationship between plants and insects. As insects gather nectar from different plants of the same species, they also spread pollen from plants on which they have previously fed. This greatly increases plants' ability to cross-pollinate, which maintains and possibly even improves their evolutionary fitness. This ultimately affects humans since ensuring healthy crops is critical to agriculture. A serious environmental problem is the decline of populations of pollinator insects, and a number of species of insects are now cultured primarily for pollination management in order to have sufficient pollinators in the field, orchard or greenhouse at bloom time. Insects also produce useful substances such as honey, wax, lacquer and silk. Honey bees have been cultured by humans for thousands of years for honey, although contracting for crop pollination is becoming more significant for beekeepers. The silkworm has greatly affected human history, as silk-driven trade established relationships between China and the rest of the world. Insects play important roles in biological research. For example, because of its small size, short generation time and high fecundity, the common fruit fly *Drosophila melanogaster* is a model organism for studies in the genetics of higher eukaryotes. *D. melanogaster* has been an essential part of studies into principles like genetic linkage, interactions between genes, chromosomal genetics, development, behaviour, and evolution.

Because genetic systems are well conserved among eukaryotes, understanding basic cellular processes like DNA replication or transcription in fruit flies can help to understand those processes in other eukaryotes, including humans. The genome of *D. melanogaster* was sequenced in 2000, reflecting the organism's important role in biological research. Insectivorous insects, or insects which feed on other insects, are beneficial to humans because they eat insects that could cause damage to agriculture and human structures. For example, aphids feed on crops and cause problems for farmers, but ladybugs feed on aphids, and can be used as a means to get significantly reduce pest aphid populations. While birds are perhaps more visible predators of insects, insects themselves account for the vast majority of insect consumption. Without predators to keep them in check, insects can undergo almost unstoppable population explosions. Many insects, especially beetles, are scavengers that feed on dead animals and fallen trees and thereby recycle biological materials into forms found useful by other organisms. Insects are responsible for much of the process by which topsoil is created. The ancient Egyptian religion considered dung beetles sacred, and represented them as beetle-shaped amulets, or scarabs. Dung beetles have been used in countries including Australia as an agent of biological pest control to reduce the populations of pestilent flies and parasitic worms. The Australian Dung Beetle Project successfully introduced 23 species of dung beetle, including *Onthophagus gazella* and *Euoniticellus intermedius* from South Africa and Europe. This resulting in a 90% reduction in bush flies as well as improved soil fertility and quality. Insects are also used in medicine, for example fly larvae (maggots) were formerly used to treat wounds to prevent or stop gangrene, as they would only consume dead flesh. This treatment is finding modern usage in some hospitals. Recently insects have also gained attention as potential sources of drugs and other medicinal substances. Adult insects, such as crickets, and insect larvae of various kinds are also commonly used as fishing bait.

Entomophagy

In some parts of the world, insects are used for human food, while being a taboo in other places. In some cultures, insects, especially deep-fried cicadas, are considered to be delicacies, while in other places they form part of the normal diet as they have a high protein content for their mass. In most first-world countries, however, entomophagy, or the consumption of insects, is taboo. There are proponents of developing this use to provide a major source of protein in human nutrition. Since it is impossible to entirely eliminate pest insects from the human food chain, insects are present in many foods, especially grains. Food safety laws in many countries do not prohibit insect parts in food, but rather limit the quantity. According to cultural materialist anthropologist Marvin Harris, the eating of insects is taboo in cultures that have other protein sources such as fish or livestock.

In Culture

Scarab beetles held religious and cultural symbolism in Old Egypt, Greece and some shamanistic Old World cultures. The ancient Chinese regarded cicadas as symbols of rebirth or immortality. In Mesopotamian literature, the epic poem of Gilgamesh has allusions to Odonata which signify the impossibility of immortality. Amongst the Aborigines of Australia of the Arrernte language groups, honey ants and witchety grubs served as personal clan totems. In the case of the 'San' bush-men of the Kalahari, it is the praying mantis which holds much cultural significance including creation and zen-like patience in waiting.

INSECT ECOLOGY

Insect ecology is the scientific study of how insects,

individually or as a community, interact with the surrounding environment or ecosystem. Insects play significant roles in the ecology of the world due to their vast diversity of form, function and life-style; their considerable biomass; and their interaction with plant life, other organisms and the environment. Since they are the major contributor to biodiversity in the majority of habitats, except in the sea, they accordingly play a variety of extremely important ecological roles in the many functions of an ecosystem. Taking the case of nutrient recycling; insects contribute to this vital function by degrading or consuming leaf litter, wood, carrion and dung and by dispersal of fungi. Insects form an important part of the food chain, especially for entomophagous vertebrates such as many mammals, birds, amphibians and reptiles. Insects play an important role in maintaining community structure and composition; in the case of animals by transmission of diseases, predation and parasitism, and in the case of plants, through phytophagy and by plant propagation through pollination and seed dispersal. From an anthropocentric point of view, insects compete with humans; they consume as much as 10% of the food produced by man and infect one in six humans with a pathogen.

INSECT WINTER ECOLOGY

Insect winter ecology entails the overwinter survival strategies of insects, which are in many respects more similar to those of plants than to many other animals, such as mammals and birds. This is because unlike those animals, which can generate their own heat internally (endothermic), insects must rely on external sources to provide their heat (ectothermic). Thus, insects sticking around in the winter, must tolerate freezing or rely on their surroundings to provide enough heat to avoid freezing. Loss of enzymatic function and eventual freezing due to low temperatures daily threatens the livelihood of these organisms during winter. Not surprisingly, insects have evolved a number of strategies to deal with the rigors of winter temperatures in places where they would otherwise not survive.

Survival Strategies

Two major strategies for winter survival have evolved in the Class Insecta due to their inability to generate significant heat metabolically. The first, migration, is a complete avoidance of the temperatures that pose a threat. If an insect cannot migrate, then it must stay and deal with the cold temperatures in one of two ways. This cold hardiness is separated into two categories, freeze avoidance and freeze tolerance.

Migration

Migration in insects is different than in birds. Bird migration is a two-way, round-trip movement of each individual, whereas this is not usually the case with insects. The short lifespan of insects compared to birds means that the adult that made one leg of the trip will be replaced by a member of the next generation on the return voyage. As a result, invertebrate biologists have redefined migration for this group of organisms as consisting of three parts:

1. A persistent, straight line movement away from the natal area
2. Distinctive pre- and post-movement behaviours
3. Re-allocation of energy within the body associated with the movement

This definition allows for mass insect movements to be considered as migration. Perhaps the best known insect migration is that of the monarch butterfly. The monarch in North America migrates from as far north as Canada southward to Mexico and Southern California annually from about August to October. The population east of the Rocky Mountains overwinters in Michoacán, Mexico, and the western population overwinters in various sites in central coastal California, notably in Pacific Grove and Santa Cruz. The round trip journey is typically around

3600 km in length. The longest one-way flight on record for monarchs is an astonishing 3009 km from Ontario, Canada to San Luis Potosí, Mexico. They use the direction of sunlight and magnetic cues to orient themselves during migration. The monarch requires significant energy to make such a long flight, which is provided by fat reserves. When they reach their overwintering sites, they begin a period of lowered metabolic rate. Nectar from flowers procured at the overwintering site provides energy for the northward migration. To limit their energy use, monarchs congregate in large clusters in order to maintain a suitable temperature. This strategy, similar to huddling in small mammals, makes use of body heat from all the organisms and lowers heat loss. Another common winter migrant insect, found in much of North America, South America, and the Caribbean, is the Green Darner. Migration patterns in this species are much less studied than those of monarchs. Green darners leave their northern ranges in September and migrate south. Studies have noted a seasonal influx of green darners to southern Florida, which indicates migratory behaviour. Little has been done with tracking of the green darner, and reasons for migration are not fully understood since there are both resident and migrant populations. The common cue for migration southward in this species is the onset of winter.

Freeze Avoidance

Lethal freezing occurs when insects are exposed to temperatures below the melting point (MP) of their body fluids; therefore, insects that do not migrate from regions with the onset of colder temperatures must either devise strategies to tolerate or avoid freezing of intracellular and extracellular body fluids. Surviving colder temperatures, in insects, generally falls under two categories: Freeze-tolerant insects can tolerate the formation of internal ice and freeze-avoidant insects avoid freezing by keeping the bodily fluids liquid. The general strategy adopted by insects also differs between the northern hemisphere and the southern hemisphere. In the northern hemisphere where cold temperatures are expected seasonally and are usually for long periods of time, the main strategy is freeze avoidance. In the southern hemisphere, where seasonal cold temperatures are not as extreme or long lasting, the main strategy is freeze tolerance. Freeze avoidance involves both physiological and biochemical mechanisms. One method of freeze avoidance is the selection of a dry hibernation site in which no ice nucleation from an external source can occur. Insects may also have a physical barrier such as a wax-coated cuticle that provides protection against external ice across the cuticle. The stage of development at which an insect over-winters varies across species, but can occur at any point of the life cycle (i.e., egg, pupa, larva, and adult).

Freeze-avoidant insects that cannot tolerate the formation of ice within their bodily fluids need to implement strategies to depress the temperature at which their bodily fluids will freeze. Supercooling is the process by which water cools below its freezing point without changing phase into a solid, due to the lack of a nucleation source. Water requires a particle such as dust in order to crystallize and if no source of nucleation is introduced, water can cool down to -42°C without freezing. In the initial phase of seasonal cold hardening, ice-nucleating agents (INAs) such as food particles, dust particles and bacteria, in the gut or intracellular compartments of freeze avoidant insects have to be removed or inactivated. Removal of ice-nucleating material from the gut can be achieved by cessation in feeding, clearing the gut and removing lipoprotein ice nucleators (LPINs) from the heamolymph and in some species, by the shedding of the mid-gut during moulting. In addition to physical preparations for winter, many insects also alter their biochemistry and metabolism. For example, some insects synthesize cryoprotectants such as polyols and sugars, which reduce the lethal freezing temperature of the body. Although polyols such as sorbitol, mannitol, and ethylene glycol can also be found, glycerol is by far the most common cryoprotectant and can be equivalent to ~20% of the total body mass. Glycerol is distributed uniformly throughout the head, the thorax, and the abdomen of insects, and is in equal concentration in intracellular and extracellular compartments. The depressive effect of glycerol on the super cooling point (SCP) is thought to be due to the high viscosity of glycerol solutions at low temperatures. This would inhibit INA activity and SCPs would drop far below the environmental temperature.

At colder temperatures (below 0 °C), glycogen production is inhibited, and the breakdown of glycogen into glycerol is enhanced, resulting in the glycerol levels in freeze avoidant insects reaching levels five times higher than those in freeze tolerant insects which do not need to cope with extended periods of cold temperatures. Though not all freeze avoidant insects produce polyols, all hibernating insects produce thermal hysteresis factors (THFs). A seasonal photoperiodic timing mechanism is responsible for increasing the antifreeze protein levels with concentrations reaching their highest in the winter. In the pyrochroid beetle, "Dendroides canadensis", a short photoperiod of 8 hours light and 16 hours of darkness, results in the highest levels of THFs, which corresponds with the shortening of daylight hours associated with winter. These antifreeze proteins are thought to stabilize SCPs by binding directly to the surface structures of the ice crystals themselves, diminishing crystal size and growth. Therefore, instead of acting to change the biochemistry of the bodily fluids as seen with cryoprotectants, THFs act directly with the ice crystals by adsorbing to the developing crystals to inhibit their growth and reduce the chance of lethal freezing occurring.

Freeze Tolerance

Freeze tolerance in insects refers to the ability of some insect species to survive ice formation within their tissues. All insects are ectothermic, which can make them vulnerable to freezing. In most animals, intra- and extracellular freezing causes severe tissue damage, resulting in death. Insects that have evolved freeze-tolerance strategies manage to avoid tissue damage by controlling where, when, and to what extent ice forms. In contrast to freeze avoiding insects that are able to exist in cold conditions by supercooling, freeze tolerant organisms limit supercooling and initiate the freezing of their body fluids at relatively high temperatures. Physiologically, this is accomplished through inoculative freezing, the production of ice nucleating proteins, crystalloid compounds, and/or microbes.

Although freeze-avoidance strategies predominate in the insects, freeze tolerance has evolved at least six times within this group (in the Lepidoptera, Blattaria, Diptera, Orthoptera, Coleoptera, and Hymenoptera). Freeze tolerance is also more prevalent in insects from the Southern Hemisphere (reported in 85% of species studied) than it is in insects from the Northern Hemisphere (reported in 29% of species studied). It has been suggested that this may be due to the Southern Hemisphere's greater climate variability, where insects must be able to survive sudden cold snaps yet take advantage of unseasonably warm weather as well. This is in contrast to the Northern Hemisphere, where predictable weather makes it more advantageous to overwinter after extensive seasonal cold hardening. Examples of freeze tolerant insects include: the woolly bear *Pyrrharctia isabella*, the flightless midge *Belgica antarctica* and the alpine cockroach *Celatoblatta quinquemaculata*.

Dangers of Freezing

With some exceptions, the formation of ice within cells generally causes cell death even in freeze-tolerant species due to physical stresses exerted as ice crystals expand. Ice formation in extracellular spaces is also problematic, as it removes water from solution through the process of osmosis, causing the cellular environment to become hypertonic and draw water from the cell interiors. Excessive cell shrinkage can cause severe damage. This is because as ice forms outside the cell, the possible shapes that can be assumed by the cells are increasingly limited, causing damaging deformation. Finally, the expansion of ice within vessels and other spaces can cause physical damage to structures and tissues.

Ice Nucleators

In order for a body of water to freeze, a nucleus must be present upon which an ice crystal can begin to grow. At low temperatures, nuclei may arise spontaneously from clusters of slow-moving water molecules. Alternatively, substances that facilitate the aggregation of water molecules can increase the probability that they will reach the critical size necessary for ice formation. Freeze-tolerant insects are known to produce ice nucleating proteins. The regulated production of ice nucleating proteins allows insects to control the formation of ice crystals within their bodies. The lower an insects' body temperature, the more likely it is that ice will begin to form spontaneously. Even freeze-tolerant animals cannot tolerate a sudden, total freeze; for most freeze-tolerant insects it is important that they avoid supercooling and initiate ice formation at relatively warm temperatures. This allows the insect to moderate the rate of ice growth, adjust more slowly to the mechanical and osmotic pressures imposed by ice formation. Nucleating proteins may be produced by the insect, or by microorganisms that have become associated with the insects' tissues. These microorganisms possess proteins within their cell walls that function as nuclei for ice growth. The temperature that a particular ice nucleator initiates freezing varies from molecule to molecule. Although an organism may possess a number of different ice nucleating proteins, only those that initiate freezing at the highest temperature will catalyze an ice nucleation event. Once freezing is initiated, ice will spread throughout the insect's body.

Cryoprotectants

The formation of ice in the extracellular fluid causes

an overall movement of water out of cells, a phenomenon known as osmosis. As too much dehydration can be dangerous to cells, many insects possess high concentrations of solutes such as glycerol. Glycerol is a relatively polar molecule and therefore attracts water molecules, shifting the osmotic balance and holding some water inside the cells. As a result, cryoprotectants like glycerol decrease the amount of ice that forms outside of cells and reduce cellular dehydration. Insect cryoprotectants are also important for species that avoid freezing; see description above

Intracellular Freezing

Most freeze-tolerant species restrict ice formation to extracellular spaces. Some species, however, can tolerate intracellular freezing as well. This was first discovered in the fat body cells of the goldenrod gall fly *Eurosta solidaginis*. The fat body is an insect tissue that is important for lipid, protein and carbohydrate metabolism (analogous to the mammalian liver). Although it is not certain why intracellular freezing is restricted to the fat body tissue in some insects, there is evidence that it may be due to the low water content within fat body cells.

Locations of Hibernating Insects

Insects are well hidden in winter, but there are several locations in which they can reliably be found. Ladybugs practice communal hibernation by stacking one on top of one another on stumps and under rocks to share heat and buffer themselves against winter temperatures. The female grasshopper (family Tettigoniidae [long-horned]), in an attempt to keep her eggs safe through the winter, tunnels into the soil and deposits her eggs as deep as possible in the ground. Many other insects, including various butterflies and moths also overwinter in soil in the egg stage. Some adult beetles hibernate underground during winter; many flies overwinter in the soil as pupae. Other methods of hibernation include the inhabitance of bark, where insects nest more toward the southern side of the tree for heat provided by the sun. Cocoons, galls, and parasitism are also common methods of hibernation.

Aquatic Insects

Insects that live under the water have different strategies for dealing with freezing than do terrestrial insects. Many insect species survive winter not as adults on land, but as larvae underneath the surface of the water. Under the water many benthic invertebrates will experience some subfreezing temperatures, especially in small streams. Aquatic insects have developed freeze tolerance much like their terrestrial counterparts. However, freeze avoidance is not an option for aquatic insects as the presence of ice in their surroundings may cause ice nucleation in their tissues. Aquatic insects have supercooling points typically around – 3° to – 7°C. In addition to using freeze tolerance, many aquatic insects migrate deeper into the water body where the temperatures are higher than at the surface. Insects such as stoneflies, mayflies, caddisflies, and dragonflies are common overwintering aquatic insects. The dance fly larvae have the lowest reported supercooling point for an aquatic insect at – 22°C.

INSECTARY PLANTS

Insectary plants is a term used to describe plants that attract insects. As such, beneficial insectary plants are intentionally introduced into an ecosystem to increase pollen resources and nectar resources required by the natural enemies of harmful or unwanted insect pests. Beyond an effective natural control of pests, the *friendly insects* also assist in pollination. The "friendly insects" include ladybeetles, bees, ground beetles, hoverflies, and parasitic wasps. Other animals that are frequently considered beneficial include lizards, spiders, toads, and humming birds. Beneficial insects are as much as ten times more abundant in the insectary plantings area. Mortality of scale insects (caused by natural enemies) can be double with insectary plantings. In addition, a diversity of insectary plants can increase the population of beneficial insects such that these levels can be sustained even when the insectary plants are removed or die off. For maximum benefit in the garden, insectary plants can be grown alongside desired garden plants that do not have this benefit. The insects attracted to the insectary plants will also help the other nearby garden plants. Many members of the Apiaceae (formerly known as Umbelliferae) family are excellent insectary plants. Fennel, angelica, coriander (cilantro), dill, and wild carrot all provide in great number the tiny flowers required by parasitic wasps. Various clovers, yarrow, and rue also attract parasitic and predatory insects.

Low-growing plants, such as thyme, rosemary, or mint, provide shelter for ground beetles and other beneficial insects. Composite flowers (daisy and chamomile) and mints (spearmint, peppermint, or catnip) will attract predatory wasps, hoverflies, and robber flies. The wasps will catch caterpillars and grubs to feed their young, while the predatory and parasitic flies attack many kinds of insects, including leafhoppers and caterpillars. Other insectary plants include: mustard (Brassica juncea), phacelia (Phacelia tanacetifolia), buckwheat (Fagopyrum esceulentum), marigold (Tagetes patula), elderberry (sambucus mexicana), and Korean licorice mint (Agastache rugosa).

INSECTIVORE

An insectivore is a type of carnivore with a diet that consists chiefly of insects and similar small creatures. An alternate term is entomophage, which also refers to the human practice of eating insects. Amphibians were the first vertebrate insectivores. When they evolved, 400 million years ago, amphibians were piscivores, with numerous sharp conical teeth, not unlike a modern crocodile. The same tooth arrangement is however also suited for eating animals with exoskeletons, thus the ability to eat insects is an extension of piscivory.

Overview of Insects

Although individually small, insects exist in enormous numbers - they number over a million described species and make up a very large part of the animal biomass in almost all non-marine environments. It has been estimated that the global insect biomass is in the region of 10 kg with an estimated population of 10 organisms. Many creatures depend on insects as their primary diet, and many that do not (and are thus not technically insectivores) nevertheless use insects as a protein supplement, particularly when they are breeding. Examples of insectivores include nightingales, aardwolfs, echidnas, swallows, anteaters, carp, frogs, lizards, bats, and spiders. Even large mammals are recorded as eating insects; the sloth bear is perhaps the largest insectivore. Insects also can be insectivores; examples are dragonflies, hornets, ladybugs, and praying mantises. Insectivory also features to various degrees amongst primates, such as marmosets, tamarins, tarsiers, galagos and aye-aye. There is some suggestion that the earliest primates were nocturnal, arboreal insectivores.

Insectivorous Plants

Insectivorous plants are plants that derive some or most of their nutrients (but not energy) from trapping and consuming animals or protozoans, typically insects and other arthropods. Also known as *carnivorous plants*, they appear adapted to grow in places where the soil is thin or poor in nutrients, especially nitrogen, such as acidic bogs and rock outcroppings. Insectivorous plants include the Venus flytrap, several types of pitcher plants, butterworts, sundews, bladderworts, the waterwheel plant, brocchinia bromeliads, and others. These plants obtain nitrogen by trapping insects. They use a variety of mechanism such as pitfalls, sticky surfaces, hair-trigger snaps, bladder-trap and lobster-pot trap mechanisms. Technically these plants are not strictly insectivorous, as they consume any animal small enough to be trapped by them; indeed, the larger varieties of pitcher plant have been known to consume small rodents and lizards. Charles Darwin wrote the first well-known treatise on carnivorous plants in 1875.

REFERENCES

Alex Rasnitsyn & D.L.J. Quicke (2002). *History of Insects.* Kluwer Academic Publishers.

Capinera, John L. (Editor). (2008). *Encyclopedia of Entomology,* (2nd ed). Springer Reference.

Chapman R.F. (1998). The Insects Structure and Function 4th ed. Cambridge University Press 520-522.

Chapman, A.D. (2006). *Numbers of living species in Australia and the World.* Canberra: Australian Biological Resources Study, p. 60.

Chown, S.L.; S.W. Nicholson (2004). *Insect Physiological Ecology.* New York: Oxford University Press.

Costanzo JP and Lee R.E. (1996). Ice nucleation in freeze-tolerant vertebrates. Cryo-Letters 17:111-118.

David Grimaldi & Michael S. Engel (2005). *Evolution of the Insects.* Cambridge University Press, p. 143.

Dudley, Robert (2002). "Flight and the Pterygote Insecta". *The biomechanics of insect flight: form, function, evolution.* Princeton University Press, pp. 3–35.

Engel, Michael S.; David A. Grimaldi (2004). "New light shed on the oldest insect". *Nature* 427 (6975): 627–630.

Gullan, P.J.; Cranston, P.S. (2005). *The insects: an outline of entomology* (3 (illustrated, revised) ed.). Wiley-Blackwell, p. 505.

Hill, Dennis S. (1997). *The economic importance of insects.* Springer, p. 198.

Horwath K.L. and Duman J.G. (1982). Involvement of the Circadian System in Photoperiodic Regulation of Insect Antifreeze Proteins. The Journal of Experimental Zoology 219:267-270.

Jones, S., Martin, R., & Pilbeam, D., ed (1994). *The Cambridge Encyclopedia of Human Evolution.* Cambridge: Cambridge University Press.

Lee R.E. and Costanzo J.P. (1998). Biological ice nucleation and ice distribution in cold-hardy ectothermic animals. Annual Review of Physiology. 60:55-72.

Lee, Richard E, Jr. (1989). "Insect Cold-Hardiness: To Freeze or Not to Freeze". *BioScience* 39 (5): 308–313.

Lloyd, James E.; Erin C. Gentry (2003). *The Encyclopedia of Insects*. Academic Press, pp. 115–120.

Marchand PJ and Walker L. (1996). Life in the Cold: An Introduction to Winter Ecology. Hanover, NH: University Press of New England.

Miller, George A. (2009). "WordNet - About Us." : entry on *insectivorous.*". Princeton University.

Rice, H.M.A. (1968). "Two Tertiary sawflies, (Hymenoptera - Tenthredinidae), from British Columbia". *Geological Survey of Canada* 67 (59): 1–21.

Sahney, S., Benton, M.J. & Falcon-Lang, H.J. (2010). "Rainforest collapse triggered Pennsylvanian tetrapod diversification in Euramerica" (PDF). *Geology* 38 (12): 1079–1082.

Speight, Martin R.; Hunter, Mark D.; Watt, Allan D. (1999). *Ecology of insects: concepts and applications* (Illustrated) ed.). Wiley-Blackwell, p. 350.

Szwedo, J.; Stroiński, A. (2001). "*Tainosia quisqueyae* gen. and sp. nov. from the Oligocene/Miocene Dominican amber (Hemiptera: Fulgoroidea: Nogodinidae)".

Tracey, J et al.; Wilson, RI; Laurent, G; Benzer, S (18 April 2003). "Painless, a Drosophila gene essential for nociception". *Cell* 113 (2): 261–273.

Whitney, Stephen R. & Sandelin, R. (2004). *Field Guide to the Cascades & Olympics*. The Mountaineers Books, p. 317.

3

Insect Phylogeny, Physiology, Sterility and Defense in Insects

PHYLOGENY OF INSECTS

Phylogeny or phylogenetics is the study of evolutionary relatedness among groups of organisms, which is discovered through molecular sequencing data and morphological data matrices. Insects are a highly diverse group of organisms with a worldwide distribution, thus having a complex evolutionary history. They have conquered every terrestrial environment and have many intricate interactions with a wide variety of organisms, including predatory-prey relationships. The taxonomy of insects is also included here, which is the classification, identification, and naming of species, as taxonomy is richly informed by phylogenetics, but remains methodologically and logically distinct. The evolution of insects dates back to the Devonian period, with the oldest definitive insect fossil being the *Rhyniognatha hirsti*, estimated at 407 to 396 million years ago. Global climate conditions changed several times during the history of the earth, along with it the diversity of insects.

The Pterygotes underwent a major radiation in the Carboniferous while the Endopterygota species underwent another major radiation in the Permian. Survivors of the mass extinction at the PT boundary evolved in the Triassic to what are essentially the modern Insecta Orders that persist to modern times. Most modern insect families appeared in the Jurassic, and further diversity probably in genera occurred in the Cretaceous. It is believed that by the Tertiary, there existed many of what are still modern genera; hence, most insects in amber are, indeed, members of extant genera. What seems most fascinating is that insects diversified in a brief 100 million years (give or take) into the modern forms that exist with minor change in modern times.

Fossils

Insect fossils are not just merely impressions, but also many appear in many other forms; While wings are indeed a common insect fossil, they do not readily decay or digest, which is why birds and spiders typically leave the wings after devouring the rest of an insect. Terrestrial vertebrates are almost always preserved just as bony remains (or inorganic casts thereof), the original bone usually having been replaced by the mineral apatite. Occasionally, mummified or frozen vertebrates are found, but their age is usually no more than several thousand years. Fossils of insects, in contrast, are preserved as three-dimensional, permineralized, and charcoalified replicas; and as inclusions in amber and even within some minerals. There is also abundant fossil evidence for the behaviour of extinct insects, including feeding damage on fossil vegetation and in wood, fecal pellets, and nests in fossil soils. Dinosaur behaviour, by contrast, is recorded mostly as footprints and coprolites.

The common denominator among most deposits of fossil insects and terrestrial plants is the lake environment. Those insects that became preserved were either living in the fossil lake (autochthonous) or carried into it from surrounding habitats by winds, stream currents, or their own flight (allochthonous). Drowning and dying insects not eaten by fish and other predators settle to the bottom, where they may be preserved in the lake's sediments, called lacustrine, under appropriate conditions. Even amber, or fossil resin from trees, requires a watery environment that is lacustrine or brackish in order to be preserved. Without protection in anoxic sediments, amber would gradually disintegrate; it is never found buried in fossil soils. Various factors contribute greatly to what kinds of insects become preserved and how well, if indeed at all, including lake depth, temperature, and alkalinity; type of sediments; whether the lake was surrounded by forest or vast and featureless salt pans; and if it was choked in anoxia or highly oxygenated. There are some major exceptions to the lacustrine theme of fossil insects, the most famous being the Late Jurassic limestones from Solnhofen and Eichstätt, Germany, which are marine. These deposits are famous for pterosaurs and the earliest bird, Archaeopteryx. The limestones were formed by a very fine mud of calcite that settled within stagnant, hypersaline bays isolated from inland seas. Most organisms in these limestones, including rare insects, were preserved intact, sometimes with feathers and outlines of soft wing membranes, indicating that there was very little decay. The insects, however, are like casts or molds, having relief but little detail. In some cases iron oxides precipitated around wing veins, revealing better detail.

There are many different ways insects can be fossilized and preserved, one which includes compressions and impressions, concretions, mineral replication, charcoalified (fusainized) remains, and their trace remains. Compressions and Impressions are the most extensive types of insect fossils, occurring in rocks from the Carboniferous to Recent. Impressions are like a cast or mold of a fossil insect, showing its form and even some relief, like pleating in the wings, but usually little or no color from the cuticle. Compressions preserve remains of the cuticle, so color exceptional situations microscopic features such as microtrichia on sclerites and wing membranes are even visible, but preservation of this scale also requires a matrix of exceptionally fine grain, such as in micritic muds and volcanic tuffs. Because arthropod sclerites are held together by membranes, which readily decompose, many fossil arthropods are known only by isolated sclerites. Far more desirable are complete fossils. Concretions are stones with a fossil at the core whose chemical composition differs from that of the surrounding matrix, usually formed as a result of mineral precipitation from decaying organisms. The most significant deposit consists of various localities of the Late Carboniferous Francis Creek Shale of the Carbondale Formation at Mazon Creek, Illinois, which are composed of shales and coal seams yielding oblong concretions. Within most concretions is a mold of an animal and sometimes a plant that is usually marine in origin. When an insect is partly or wholly replaced by minerals, usually completely articulated and with three-dimensional fidelity, is called Mineral replication. This is also called petrifaction, as in "petrified" wood. Insects preserved this way are often, but not always, preserved as concretions, or within nodules of minerals that formed around the insect as its nucleus. Such deposits generally form where the sediments and water are laden with minerals, and where there is also quick mineralization of the carcass by coats of bacteria. Encapsulation Charcoalified (fusainized) remains trace remains

Evolutionary History

Global climate conditions changed several times during the history of the earth, along with it the diversity of insects. They can be reconstructed until about 2,300 million years ago by sediment analysis. During the Phanerozoic several ice ages occurred, characterized by a complete glaciation of the polar caps. From a geological point of view, today we live in an ice age as well. However, this is not at all the "normal" climate condition of the earth, on the contrary, during the last 550 mio. years acryogenic warm periods prevailed for around 80% of the time, with the polar caps completely free of ice. Just as the Paleozoic is often called the age of the trilobite, modern time is often called the age of the insect. The insect fossil record extends back some 400 million years to the lower Devonian, while the Pterygotes (winged insects) underwent a major radiation in the Carboniferous. The Endopterygota underwent a another major radiation in the Permian. Survivors of the mass extinction at the PT boundary evolved in the Triassic to what are essentially the modern Insecta Orders that persist to modern times. Most modern insect families appeared in the Jurassic, and further diversity probably in genera occurred in the Cretaceous. It is believed that by the Tertiary, there existed many of what are still modern genera; hence, most insects in amber are, indeed, members of extant genera. What seems most fascinating is that insects diversified in a brief 100 million years into the modern forms that exist with minor change in modern times.

Insect evolution is characterized by rapid adaptation with selective pressures exerted by environment, with rapid adaptation being furthered by their high fecundity. It appears the rapid radiation's, and to this day the appearance of new species result in insects filling all available environmental niches. Insect evolution is closely related to the evolution of flowering plants. Insect adaptations include feeding on flowers and related structures with some 20% of extant insects depend on flowers, nectar or pollen for their food source. This symbiotic relationship is even more paramount in evolution considering that about 2/3 of flowering plants are insect pollinated. Insects are also vectors of many pathogens that may even have been responsible for the decimation or extinction of some mammalian species. Compared to other organisms, insects have not left a particularly robust fossil record. Other than in amber, most insects are terrestrial and only preserved under very special conditions such as at the edge of freshwater lakes. Yet in amber, age is limited since large resin production by trees developed later than the ancient insects. Interestingly, while some 1/3 of non-insect species are extinct fossils, it is believed that only 1/100th insects are extinct fossils.

Devonian

The Devonian (408 to 362 million years ago) was a

relatively warm period, and probably lacked any glaciers with reconstruction of tropical sea surface temperature from conodont apatite implies an average value of 30 °C (86 °F) in the Early Devonian. CO_2 levels dropped steeply throughout the Devonian period as the burial of the newly-evolved forests drew carbon out of the atmosphere into sediments; this may be reflected by a Mid-Devonian cooling of around 5 °C (9 °F). The Late Devonian warmed to levels equivalent to the Early Devonian; while there is no corresponding increase in CO_2 concentrations, continental weathering increases (as predicted by warmer temperatures); further, a range of evidence, such as plant distribution, points to Late Devonian warming. The continent Euramerica (or Laurussia) was created in the early Devonian by the collision of Laurentia and Baltica, which rotated into the natural dry zone along the Tropic of Capricorn, which is formed as much in Paleozoic times as nowadays by the convergence of two great air-masses, the Hadley cell and the Ferrel cell. The oldest definitive insect fossil is the Devonian *Rhyniognatha hirsti*, estimated at 396-407 million years old. This species already possessed dicondylic mandibles, a feature associated with winged insects, suggesting that wings may already have evolved at this time. Thus, the first insects probably appeared earlier, in the Silurian period. Like other insects of its time, *Rhyniognatha* presumably fed on plant sporophylls - which occur at the tips of branches and bear sporangia, the spore-producing organs. The insect's anatomy might also give clues as to what it ate. The creature had large mandibles which may or may not have been used for hunting.

Carboniferous

The Carboniferous 362 to 290 million years ago) is famous for its wet, warm climates and extensive swamps of mosses, ferns, horsetails, and calamites. Glaciations in Gondwana, triggered by Gondwana's southward movement, continued into the Permian and because of the lack of clear markers and breaks, the deposits of this glacial period are often referred to as Permo-Carboniferous in age. The cooling and drying of the climate led to the Carboniferous Rainforest Collapse (CRC). Tropical rain forests fragmented and then were eventually devastated by climate change. Remains of insects are scattered throughout the coal deposits, particularly of wings from cockroaches (Blattodea); two deposits in particular are from Mazon Creek, Illinois and Commentary, France. The earliest winged insects are from this time period (Pterygota), including the aforementioned Blattaria, Caloneurodea, primitive stem-group Ephemeropterans, Orthoptera, Palaeodictyopteroidea. In 1940, in Noble County, Oklahoma fossil of Meganeuropsis americana as it represented by the largest complete insect wing ever found.

Very early Blattarians had a large, discoid pronotum and coriaceous forewings wings with a distinct CuP vein (a unbranched wing vein, lying near the claval fold and reaching the wing posterior margin). These were not true cockroaches, as they had an ovipositor, although through the Carboniferous, the ovipositor started to diminish. The orders Caloneurodea and Miomoptera are known, with orthoptera and Blattaria to be one of the earliest Neoptera; developing from the upper Caboniferous to the Permian. These insects had wings with similar form and structure: small anal lobes. Species of Orthoptera, or grasshopers and related kin, is an ancient order that still exist till today extending from this time period. From which time even the distinctive synapomorphy of saltatorial, or adaptive for jumping, hind legs is preserved. Palaeodictyopteroidea is a large and diverse group includes 50% of all known Paleozoic insects. Containing many of the primitive features of the time: very long cerci, an ovipositor, and wings with little or no anal lobe. Protodonata, as its name implies, is a primitive paraphyletic group similar to Odonata; although lacks distinct features such as a nodus, a pterostigma and a arculus. Most were only slightly larger than modern dragonflies, but the group does include the largest known insects, such as the late Carboniferous *Meganeura monyi*,

Megatypus, and the even larger later Permian *Meganeuropsis permiana*, with wingspans of up to 71 centimetres (2.33 ft). They were probably the top predators for some 100 million years.

Permian

The Permian (290 to 245 million years ago) was a relatively short period, however a very important period at least. During this time period, all the Earth's major land masses were collected into a single supercontinent known as Pangaea. Pangaea straddled the equator and extended toward the poles, with a corresponding effect on ocean currents in the single great ocean ("Panthalassa", the "universal sea"), and the Paleo-Tethys Ocean, a large ocean that was between Asia and Gondwana. The Cimmeria continent rifted away from Gondwana and drifted north to Laurasia, causing the Paleo-Tethys to shrink. At the end of the Permian, the biggest mass extinction in history occurred, collectively called the Permian–Triassic extinction event: 30% of all insect species became extinct; however, this is the only known mass insect extinction in Earth's history. 2007 study based on DNA of living beetles and maps of likely beetle evolution indicated that beetles may have originated during the Lower Permian, up to 299 million years ago. In 2009, a fossil beetle was described from the Pennsylvanian of Mazon Creek, Illinois, pushing the origin of the beetles to an earlier date, 318 to 299 million years ago. Fossils from this time have been found in Asia and Europe, for instance in the red slate fossil beds of Niedermoschel near Mainz, Germany. Further fossils have been found in Obora, Czechia and Tshekarda in the Ural mountains, Russia. However, there are only a few fossils from North America before the middle Permian, although both Asia and North America had been united to Euramerica.

The first discoveries from North America were made in the Wellington formation of Oklahoma and were published in 2005 and 2008. Some of the most important fossil deposits from this era are from Elmo, Kansas (260 mya); others include New South Wales, Australia (240 mya) and central Eurasia (250 mya). During this time, many of the species from the Carboniferous diversified, and many new orders developed, including: Protelytroptera, primitive relatives of Plecoptera (Paraplecoptera), Psocoptera, Mecoptera, Coleoptera, Raphidioptera, and Neuroptera. The last four being the first definitive records of the Holometabola. By the Pennsylvanian and well into the Permian, by far the most successful were primitive Blattoptera, or relatives of cockroaches. Six fast legs, two well-developed folding wings, fairly good eyes, long, well-developed antennae (olfactory), an omnivorous digestive system, a receptacle for storing sperm, a chitin skeleton that could support and protect, as well as form of gizzard and efficient mouth parts, gave it formidable advantages over other herbivorous animals. About 90% of insects were cockroach-like insects ("Blattopterans"). The dragonflies *Odonata* were the dominant aerial predator and probably dominated terrestrial insect predation as well. True Odonata appeared in the Permian and all are amphibious.

Their prototypes are the oldest winged fossils, go back to the Devonian, and are different from other wings in every way. Their prototypes may have had the beginnings of many modern attributes even by late Carboniferous and it is possible that they even captured small vertebrates, for some species had a wing span of 71 cm. The oldest known insect that resembles species of Coleoptera date back to the Lower Permian (270 million years ago), though they instead have 13-segmented antennae, elytra with more fully developed venation and more irregular longitudinal ribbing, and an abdomen and ovipositor extending beyond the apex of the elytra. The oldest true beetle, that is having features that include 11-segmented antennae, regular longitudinal ribbing on the elytra, and having genitalia that are internal. The earliest beetle-like species had pointed, leather like forewings with cells and pits. Hemiptera, or true bugs had appeared in the form of

Arctiniscytina and Paraknightia. The later had expanded parapronotal lobes, a large ovipositor, and forewings with unusual venation, possibly diverging from Blattoptera. The orders Raphidioptera and Neuroptera are grouped together as Neuropterida. The one family of putative Raphidiopteran clade (Sojanoraphidiidae) has been controversially placed as so. Although the group had a long ovipositor distinctive to this order and a series of short crossveins, however with a primitive wing venation. Early families of Plecoptera had wing venation consistent with the order and its recent decendents. Psocoptera was first appeared in the Permian period, they are often regarded as the most primitive of the hemipteroids.

Triassic

The Triassic (245 to 208 million years ago) was a period when arid and semiarid savannas developed and when the first mammals, dinosaurs, and pterosaurs also appeared. During the Triassic, almost all the Earth's land mass was still concentrated into Pangaea. From the east a vast gulf entered Pangaea, the Tethys sea. The remaining shores were surrounded by the world-ocean known as Panthalassa. The supercontinent Pangaea was rifting during the Triassic—especially late in the period—but had not yet separated. The climate of the Triassic was generally hot and dry, forming typical red bed sandstones and evaporites. There is no evidence of glaciation at or near either pole; in fact, the polar regions were apparently moist and temperate, a climate suitable for reptile-like creatures. Pangaea's large size limited the moderating effect of the global ocean; its continental climate was highly seasonal, with very hot summers and cold winters. It probably had strong, cross-equatorial monsoons.

As a consequence of the P-Tr Mass Extinction at the border of Permian and Triassic, there is only little fossil record of insects including beetles from the Lower Triassic. However, there are a few exemptions, like in Eastern Europe: At the Babiy Kamen site in the Kuznetsk Basin numerous beetle fossils were discovered, even entire specimen of the infraorders Archostemata (e.i., Ademosynidae, Schizocoleidae), Adephaga (e.i., Triaplidae, Trachypachidae) and Polyphaga (e.i., Hydrophilidae, Byrrhidae, Elateroidea) and in nearly a perfectly preserved condition. However, species from the families Cupedidae and Schizophoroidae are not present at this site, whereas they dominate at other fossil sites from the Lower Triassic. Further records are known from Khey-Yaga, Russia in the Korotaikha Basin. There are many important sites from the Jurassic, with more than 150 important sites with beetle fossils, the majority being situated in Eastern Europe and North Asia. In North America and especially in South America and Africa the number of sites from that time period is smaller and the sites have not been exhaustively investigated yet. Outstanding fossil sites include Solnhofen in Upper Bavaria, Germany, Karatau in South Kazakhstan, the Yixian formation in Liaoning, North China as well as the Jiulongshan formation and further fossil sites in Mongolia. In North America there are only a few sites with fossil records of insects from the Jurassic, namely the shell limestone deposits in the Hartford basin, the Deerfield basin and the Newark basin.

Around this time, during the Late Triassic, mycetophagous, or fungus feeding species of beetle (e.i. Cupedidae) appear in the fossil record. In the stages of the Upper Triassic representatives of the algophagous, or algae feeding species (e.i. Triaplidae and Hydrophilidae) begin to appear, as well as predatory water beetles. The first primitive weevils appear (e.i. Obrienidae), as well as the first representatives of the rove beetles (e.i. Staphylinidae), which show no marked difference in physique compared to recent species. This was also around the first time evidence of diverse freshwater insect fauna appeared. Some of the oldest living families also appear around during the Triassic, including from Hemiptera: Cercopidae, Cicadellidae, Cixiidae, and Membracidae; from Coleoptera: Carabidae, Staphylinidae, and Trachypachidae; from Hymenoptera:

Xyelidae; From Diptera: Anisopodidae, Chironomidae, and Tipulidae. The first flies (Diptera), Hymenoptera, and true dragonflies (Odonata), Heteroptera, and Thysanoptera. The first true species of Diptera are known from the Middle Triassic, becoming widespread during the Middle and Late Triassic . A single large wing from a species of Diptera in the Triassic (10 mm instead of usual 2–6 mm) was found in Australia (Mt. Crosby). This family Tilliardipteridae, despite of the numerous 'tipuloid' features, should be included in Psychodomorpha sensu Hennig on account of loss of the convex distal 1A reaching wing margin and formation of the anal loop.

Jurassic

The Jurassic (208 to 145 million years ago) was important in the development of birds, one of insects major predators. During the early Jurassic period, the supercontinent Pangaea broke up into the northern supercontinent Laurasia and the southern supercontinent Gondwana; the Gulf of Mexico opened in the new rift between North America and what is now Mexico's Yucatan Peninsula. The Jurassic North Atlantic Ocean was relatively narrow, while the South Atlantic did not open until the following Cretaceous Period, when Gondwana itself rifted apart. The global climate during the Jurassic was warm and humid. Similar to the Triassic, there were no larger landmasses situated near the polar caps and consequently, no inland ice sheets existed during the Jurassic. Although some areas of North and South America and Africa stayed arid, large parts of the continental landmasses were lush. The laurasian and the gondwanian fauna differed considerably in the Early Jurassic. Later it became more intercontinental and many species started to spread globally.

There are many important sites from the Jurassic, with more than 150 important sites with beetle fossils, the majority being situated in Eastern Europe and North Asia. In North America and especially in South America and Africa the number of sites from that time period is smaller and the sites have not been exhaustively investigated yet. Outstanding fossil sites include Solnhofen in Upper Bavaria, Germany, Karatau in South Kazakhstan, the Yixian formation in Liaoning, North China as well as the Jiulongshan formation and further fossil sites in Mongolia. In North America there are only a few sites with fossil records of insects from the Jurassic, namely the shell limestone deposits in the Hartford basin, the Deerfield basin and the Newark basin. Numerous deposits of other insects occur in Europe and Asia. Including Grimmen and Solnhofen, German; Solnhofen being famous for findings of the earliest birds (e.i., Archeopteryx). Others include Dorsect, England; Issyk-Kul, Kirghizstan; and the most productive site of all, Karatau, Kazakhstan. During the Jurassic there was a dramatic increase in the known diversity of family-level Coleoptera. This includes the development and growth of carnivorous and herbivorous species. Species of the superfamily Chrysomeloidea are believed to have developed around the same time, which include a wide array of plant host ranging from cycads and conifers, to angiosperms. Close to the Upper Jurassic, the portion of the Cupedidae decreased, however at the same time the diversity of the early plant eating, or phytophagous species increased. Most of the recent phytophagous species of Coleoptera feed on flowering plants or angiosperms.

Cretaceous

The Cretaceous (145 to 65 million years ago) had much of the same insect fauna as the Jurassic until much later on. During the Cretaceous, the late-Paleozoic-to-early-Mesozoic supercontinent of Pangaea completed its tectonic breakup into present day continents, although their positions were substantially different at the time. As the Atlantic Ocean widened, the convergent-margin orogenies that had begun during the Jurassic continued in the North American Cordillera, as the Nevadan orogeny was followed by the Sevier and Laramide orogenies. Though Gondwana was still intact in the beginning of the Cretaceous, it broke up as South America, Antarctica and Australia rifted away from Africa (though India and Madagascar remained attached to each other); thus, the South Atlantic and Indian Oceans were newly formed. Such active rifting lifted great undersea mountain chains along the welts, raising eustatic sea levels worldwide. To the north of Africa the Tethys Sea continued to narrow. Broad shallow seas advanced across central North America (the Western Interior Seaway) and Europe, then receded late in the period, leaving thick marine deposits sandwiched between coal beds. At the peak of the Cretaceous transgression, one-third of Earth's present land area was submerged. The Berriasian epoch showed a cooling trend that had been seen in the last epoch of the Jurassic. There is evidence that snowfalls were common in the higher latitudes and the tropics became wetter than during the Triassic and Jurassic. Glaciation was however restricted to alpine glaciers on some high-latitude mountains, though seasonal snow may have existed farther south. Rafting by ice of stones into marine environments occurred during much of the Cretaceous but evidence of deposition directly from glaciers is limited to the Early Cretaceous of the Eromanga Basin in southern Australia.

There is a large number of important fossil sites worldwide containing beetles from the Cretaceous. Most of them are located in Europe and Asia and belong to the temperate climate zone during the Cretaceous. A few of the fossil sites mentioned in the chapter Jurassic also shed some light on the early cretaceous beetle fauna (e.g. the Yixian formation in Liaoning, North China). Further important sites from the Lower Cretaceous include the Crato Fossil Beds in the Araripe basin in the Ceara, North Brazil as well as overlying Santana formation, with the latter was situated near the paleoequator, or the position of the earth's equator in the geologic past as defined for a specific geologic period. In Spain there are important sites near Montsec and Las Hoyas. In Australia the Koonwarra fossil beds of the Korumburra group, South Gippsland,

Victoria is noteworthy. Important fossil sites from the Upper Cretaceous are Kzyl-Dzhar in South Kazakhstan and Arkagala in Russia.

During the Cretaceous the diversity of Cupedidae and Archostemata decreased considerably. Predatory ground beetles (Carabidae) and rove beetles (Staphylinidae) began to distribute into different patterns: whereas the Carabidae predominantly occurred in the warm regions, the Staphylinidae and click beetles (Elateridae) preferred many areas with temperate climate. Likewise, predatory species of Cleroidea and Cucujoidea, hunted their prey under the bark of trees together with the jewel beetles (Buprestidae). The jewel beetles diversity increased rapidly during the Cretaceous, as they were the primary consumers of wood, while longhorn beetles (Cerambycidae) were rather rare and their diversity increased only towards the end of the Upper Cretaceous. The first coprophagous beetles have been recorded from the Upper Cretaceous, and are believed to have lived on the excrement of herbivorous dinosaurs, however there is still a discussion, whether the beetles were always tied to mammals during its development. Also, the first species with an adaption of both larvae and adults to the aquatic lifestyle are found. Whirligig beetles (Gyrinidae) were moderately diverse, although other early beetles (i.e., Dytiscidae) were less, with the most widespread being the species of Coptoclavidae, which preyed on aquatic fly larvae.

Paleogene

The Paleogene (65 to 56 million years ago) comprises the first part of the Cenozoic, which during this time the continents assumed their modern shapes. The fragments of Gondwana (South America, Africa, India and Australia) began to drift northwards. The collision of India with the eurasian landmass led to the folding and formation the Himalayas. Similarly, the Alps were folded in Central Europe by the collision of the African plate with Europe. A land bridge between North America and South America did not yet exist. The Atlantic Ocean continued to widen during the Paleogene. In the North, the last land bridge between North America and Europe broke up during the Eocene. Climate during the Paleogene was warm and tropical as most time during the Mesozoic. The climate in the beginning climate was drier and cooler than in the preceding Cretaceous, but the temperature strongly increased during the Eocene and subtropical vegetation spread up to Greenland and Patagonia. The climate near the poles was cool temperate, in Europe, North America, Australia and the southern part of South America warm temperate. Near the equator there was tropical climate, flanked by hot and arid zones in the north and the south. In the Oligocene, global cooling started. Antarctica was covered by an ice sheet and subsequently, sea levels dropped. Except an intermittent warm period during the late Oligocene, global cooling continued and finally led to the Pleistocene ice age.

There are many fossils of beetles known from this era, though the beetle fauna of the Paleocene is comparatively poorly investigated. In contrast, the knowledge on the Eocene beetle fauna is very good. The reason is the occurrence of fossil insects in amber and cay slate sediments. Amber is fossilized tree resin, that means it consists of fossilized organic compounds, not minerals. Different amber is distinguished by location, age and species of the resin producing plant. For the research on the Oligocene beetle fauna, Baltic and Dominican amber is most important. Even with the insect fossils record in general lacking, the most diverse deposit being from the Fur Formation, Denmark; including giant ants and primitive moths (Noctuidae). The first butterflies are from the Upper Paleogene, while most, like beetles, already had recent genera and species already existed during the Miocene, however, their distribution differed considerably from today's.

Neogene

The Neogene (23 to 0 million years ago), the continents

finally assumed their today's position. The South American continent drifted to the west towards the subduction zone in the Pacific, during this process the Andes were folded. During the Pliocene (5 million years ago) the land bridge between South America and North America was formed, and the fauna exchange started. The formation of this land bridge also had an impact on global climate. The Indian subcontinent continued it collision with Asia, but added an westward movement as well, leading to the folding of the Caucasus. The folding of the Himalaya continues until today. The collision of Africa with Europe and the rise of the lithosphere under the Alborán Sea (westernmost Mediterranean) lead to the separation of the Mediterranean from the Atlantic Ocean. During this period, that lasted 600,000 years (6 to 5.3 million years ago), the Mediterranean desiccated nearly completely (Messinian salinity crisis). Only at the end of this period the desiccated basin was flooded through a narrow canal near Gibraltar, according to today's view quickly, but without catastrophic effects. The Neogene was a period of global cooling, which finally led to the Pleistocene ice age. At the beginning of the Miocene, temperatures in the northern hemisphere initially were still temperate. However, by the formation of the land bridge between South America and North America, the warm ocean current was cut off and the polar caps cooled down dramatically. During the Gelasian period ice sheets began to form both in the Arctic and Antarctic region.

This marked the starting point of a new ice age which continues until today, with glacial cycles and intermittent warmer periods (interglacials). During the glacials the continental glaciers pushed to the 40th parallel in some regions and covered major parts of North America, Europe and Siberia. Each glacial advance tied up large volumes of water and the sea levels dropped globally by around 100 m. During the interglacials, the sea level rose again and coastal flooding was common during this time. The most important sites for beetle fossils of the Miocene are situated in the warm temperate and to subtropical zones. Many recent genera and species already existed during the Miocene, however, their distribution differed considerably from today's. One of the most important fossil sites for insects of the Pliocene is Willershausen near Göttingen, Germany with excellently preserved beetle fossils of various families (longhorn beetles, weevils, ladybugs and others) as well as representatives of other orders of insects. In the Willershausen clay pit so far 35 genera from 18 beetle families have been recorded, of with six genera are extinct. The Pleistocene beetle fauna is relatively well known, who used the composition of the beetle fauna to reconstruct climate conditions in the Rocky Mountains and on Beringia, the former land bridge between Asia and North America.

Phylogeny

The evolutionary relationships of insects to other animal groups remain unclear. Although more traditionally grouped with millipedes and centipedes, evidence has emerged favouring closer evolutionary ties with crustaceans. In the Pancrustacea theory, insects, together with Remipedia and Malacostraca, make up a natural clade. Other terrestrial arthropods, such as centipedes, millipedes, scorpions and spiders, are sometimes confused with insects since their body plans can appear similar, sharing (as do all arthropods) a jointed exoskeleton. However, upon closer examination their features differ significantly; most noticeably they do not have the six legs characteristic of adult insects.

The higher-level phylogeny of the arthropods continues to be a matter of debate and research. In 2008, researchers at Tufts University uncovered what they believe is the world's oldest known full-body impression of a primitive flying insect, a 300 million-year-old specimen from the Carboniferous Period. The oldest definitive insect fossil is the Devonian *Rhyniognatha hirsti*, from the 396 million year old Rhynie chert. It may have superficially resembled a modern-day silverfish insect. This species already possessed dicondylic mandibles (two articulations in the mandible), a feature associated with winged insects, suggesting that wings may already have evolved at this time. Thus, the first insects probably appeared earlier, in the Silurian period. There have been four super radiations of insects: beetles (evolved around 300 million years ago), flies (evolved around 250 million years ago), moths and wasps (evolved around 150 million years ago). These four groups account for the majority of described species. The flies and moths along with the fleas evolved from the Mecoptera. The origins of insect flight remain obscure, since the earliest winged insects currently known appear to have been capable fliers. Some extinct insects had an additional pair of winglets attaching to the first segment of the thorax, for a total of three pairs. As of 2009, there is no evidence that suggests that the insects were a particularly successful group of animals before they evolved to have wings.

Late Carboniferous and Early Permian insect orders include both extant groups and a number of Paleozoic species, now extinct. During this era, some giant dragonfly-like forms reached wingspans of 55 to 70 cm (22 to 28 in) making them far larger than any living insect. This gigantism may have been due to higher atmospheric oxygen levels that allowed increased respiratory efficiency relative to today. The lack of flying vertebrates could have been another factor. Most extinct orders of insects developed during the Permian period that began around 270 million years ago. Many of the early groups became extinct during the Permian-Triassic extinction event, the largest mass extinction in the history of the Earth, around 252 million years ago. The remarkably successful Hymenopterans appeared as long as 146 million years ago in the Cretaceous period, but achieved their wide diversity more recently in the Cenozoic era, which began 66 million years ago. A number of highly successful insect groups evolved in conjunction with flowering plants, a powerful illustration of coevolution. Many modern insect genera developed during the Cenozoic. Insects from this period on are often found preserved in amber, often in perfect condition. The body plan, or *morphology*, of such specimens is thus easily compared with modern species. The study of fossilized insects is called paleoentomology.

Evolutionary Relationships

Insects are prey for a variety of organisms, including terrestrial vertebrates. The earliest vertebrates on land existed 200 million years ago and were large amphibious piscivores, through gradual evolutionary change, insectivory was the next diet type to evolve. Insects were among the earliest terrestrial herbivores and acted as major selection agents on plants. Plants evolved chemical defenses against this herbivory and the insects in turn evolved mechanisms to deal with plant toxins. Many insects make use of these toxins to protect themselves from their predators. Such insects often advertise their toxicity using warning colors. This successful evolutionary pattern has also been utilized by mimics. Over time, this has led to complex groups of coevolved species. Conversely, some interactions between plants and insects, like pollination, are beneficial to both organisms. Coevolution has led to the development of very specific mutualisms in such systems.

Taxonomy

Traditional morphology-based or appearance-based systematics has usually given Hexapoda the rank of superclass, and identified four groups within it: insects (Ectognatha), springtails (Collembola), Protura and Diplura, the latter three being grouped together as Entognatha on the basis of internalized mouth parts. Supraordinal relationships have undergone numerous changes with the advent of methods based on evolutionary history and genetic data. A recent theory is that Hexapoda is polyphyletic (where the last common ancestor was not a member of the group), with the entognath classes having separate evolutionary histories from Insecta. Many of the traditional appearance-based taxa have been shown to be paraphyletic, so rather than using ranks like subclass, superorder and infraorder, it has proved better to use monophyletic groupings (in which the last common ancestor is a member of the group). The following represents the best supported monophyletic groupings for the Insecta. Insects can be divided into two groups historically treated as subclasses: wingless insects, known as Apterygota, and winged insects, known as Pterygota. The Apterygota consist of the primitively wingless order of the silverfish (Thysanura). Archaeognatha make up the Monocondylia based on the shape of their mandibles, while Thysanura and Pterygota are grouped together as Dicondylia. It is possible that the Thysanura themselves are not monophyletic, with the family Lepidotrichidae being a sister group to the Dicondylia (Pterygota and the remaining Thysanura).

Paleoptera and Neoptera are the winged orders of insects differentiated by the presence of hardened body parts called sclerites; also, in Neoptera, muscles that allow their wings to fold flatly over the abdomen. Neoptera can further be divided into incomplete metamorphosis-based (Polyneoptera and Paraneoptera) and complete metamorphosis-based groups. It has proved difficult to clarify the relationships between the orders in Polyneoptera because of constant new findings calling for revision of the taxa. For example, Paraneoptera has turned out to be more closely related to Endopterygota than to the rest of the Exopterygota. The recent molecular finding that the traditional louse orders Mallophaga and Anoplura are derived from within Psocoptera has led to the new taxon Psocodea. Phasmatodea and Embiidina have been suggested to form Eukinolabia. Mantodea, Blattodea and Isoptera are thought to form a monophyletic group termed Dictyoptera. It is likely that Exopterygota is paraphyletic in regard to Endopterygota. Matters that have had a lot of controversy include Strepsiptera and Diptera grouped together as Halteria based on a reduction of one of the wing pairs – a position not well-supported in the entomological community. The Neuropterida are often lumped or split on the whims of the taxonomist. Fleas are now thought to be closely related to boreid mecopterans. Many questions remain to be answered when it comes to basal relationships amongst endopterygote orders, particularly Hymenoptera. The study of the classification or taxonomy of any insect is called systematic entomology. If one works with a more specific order or even a family, the term may also be made specific to that order or family, for example systematic dipterology.

Early Evidence

The oldest definitive insect fossil is the Devonian *Rhyniognatha hirsti*, estimated at 396-407 million years old. This species already possessed dicondylic mandibles, a feature associated with winged insects, suggesting that wings may already have evolved at this time. Thus, the first insects probably appeared earlier, in the Silurian period. The subclass Apterygota (wingless insects) is now considered artificial as the silverfish (order Thysanura) are more closely related to Pterygota (winged insects) than to bristletails (order Archaeognatha). For instance, just like flying insects, Thysanura have so-called dicondylic mandibles, while Archaeognatha have monocondylic mandibles. The reason for their resemblance is not due to a particularly close relationship, but rather because they both have kept a primitive and original anatomy in a much higher degree than the winged insects. The most primitive order of flying insects, the mayflies (Ephemeroptera), are also those who are most morphologically and physiologically similar to these wingless insects. Some mayfly nymphs resemble aquatic thysanurans.

Modern Archaeognatha and Thysanura still have rudimentary appendages on their abdomen called styli, while more primitive and extinct insects known as Monura had much more developed abdominal appendages, as seen here. The abdominal and thoracic segments in the earliest terrestrial ancestor of the insects would have been more similar to each other than they are today, and the head had well-developed compound eyes and long antennae.

Their body size is not known yet. As the most primitive group today, Archaeognatha, is most abundant near the coasts, it could mean that this was the kind of habitat where the insect ancestors became terrestrial. But this specialization to coastal niches could also have a secondary origin, just as could their jumping locomotion, as it is the crawling Thysanura who are considered to be most original (plesiomorphic). By looking at how primitive cheliceratan book gills (still seen in horseshoe crabs) evolved into book lungs in primitive spiders and finally into tracheae in more advanced spiders (most of them still have a pair of book lungs intact as well), it is possible the trachea of insects was formed in a similar way, modifying gills at the base of their appendages. So far there is nothing that suggests the insects were a particularly successful group of animals before they got their wings.

Odonata

The Odonata (dragonflies) are also a good candidate as the oldest living member of the Pterygota. Mayflies are morphologically and physiologically more primitive, but the derived and advanced characteristics of dragonflies could have evolved independently in their own direction for a long time. It seems that orders with aquatic nymphs or larvae become evolutionarily conservative once they had adapted to water. If mayflies made it to the water first, this could partly explain why they are more primitive than dragonflies, even if dragonflies have an older origin. Similarly, stoneflies are the most primitive of the Neoptera, but they were not necessarily the first order to branch off. This also makes it less likely that an aquatic ancestor would have the evolutionary potential to give rise to all the different forms and species of insects that we know today. Dragonfly nymphs have a unique labial "mask" used for catching prey, and the imago has a unique way of copulating, using a secondary male sex organ on the second abdominal segment. It looks like abdominal appendages modified for sperm transfer and direct insemination have occurred at least twice in insect evolution, once in Odonata and once in the other flying insects. If these two different methods are the original ways of copulating for each group, it is a strong indication that it is the dragonflies who are the oldest, not the mayflies.

There is still not agreement about this. Another scenario is that abdominal appendages adapted for direct insemination have evolved three times in insects; once Odonata, once in mayflies and once in the Neoptera, both mayflies and Neoptera choosing the same solution. If so, it is still possible that mayflies are the oldest order among the flying insects. The power of flight is assumed to have evolved only once, suggesting sperm transfer in the earliest flying insects still was done indirectly. One possible scenario on how direct insemination evolved in insects is seen in scorpions. The male deposits a spermatophore on the ground, locks its claws with the female's claws and then guides her over his packet of sperm, making sure it comes in contact with her genital opening. When the early (male) insects laid their spermatophores on the ground, it seems likely that some of them used the clasping organs at the end of their body to drag the female over the package. The ancestors of Odonata evolved the habit of grabbing the female behind her head, as they still do today. This action, rather than not grasping the female at all, would have increased the male's chances of spreading its genes. The chances would be further increased if they first attached their spermatophore safely on their own abdomen before they placed their abdominal claspers behind the female's head; the male would then not let the female go before her abdomen had made direct contact with his sperm storage, allowing the transfer of all sperm.

This also meant increased freedom in searching for a female mate because the males could now transport the packet of sperm elsewhere if the first female slipped away. This ability would eliminate the need to either wait for another female at the site of the deposited sperm packet or to produce a new packet, wasting energy. Other advantages include the possibility of mating in other, safer places than flat ground, such as in trees or bushes. If the ancestors of the other flying insects evolved the same habit of clasping the female and dragging her over their spermathophore, but posterior instead of anterior like the Odonata does, their genitals would come very close to each others. And from there on, it would be a very short step to modify the vestigial appendages near the male genital opening to transfer the sperm directly into the female. The same appendages the male Odonata use to transfer their sperm to their secondary sexual organs at the front of their abdomen. All insects with an aquatic nymphal or larval stage seem to have adapted to water secondarily from terrestrial ancestors. Of the most primitive insects with no wings at all, Archaeognatha and Thysanura, all members live their entire life cycle in terrestrial environments. As mentioned previously, Archaeognatha were the first to split off from the branch that led to the winged insects (Pterygota), and then the Thysanura branched off.

This indicates that these three groups (Archaeognatha, Thysanura and Pterygota) have a common terrestrial ancestor, which probably resembled a primitive model of Apterygota, was an opportunistic generalist and laid spermatophores on the ground instead of copulating, like Thysanura still do today. If it had feeding habits similar to the majority of apterygotes of today, it lived mostly as a decomposer. One should expect that a gill breathing arthropod would modify its gills to breathe air if it were adapting to terrestrial environments, and not evolve new respiration organs from bottom up next to the original and still functioning ones. Then comes the fact that insect (larva and nymph) gills are actually a part of a modified, closed trachea system specially adapted for water, called tracheal gills. The arthropod trachea can only arise in an atmosphere and as a consequence of the adaptations of living on land. This too indicates that insects are descended from a terrestrial ancestor. And finally when looking at the three most primitive insects with aquatic nymphs (called naiads: Ephemeroptera, Odonata and Plecoptera), each order has its own kind of tracheal gills that are so different from one another that they must have separate origins. This would be expected if they evolved from land-dwelling species. This means that one of the most interesting parts of insect evolution is what happened between the Thysanura-Pterygota split and the first flight.

Origin of Insect Flight

The origin of insect flight remains obscure, since the earliest winged insects currently known appear to have been capable fliers. Some extinct insects (e.g. the Palaeodictyoptera) had an additional pair of winglets attached to the first segment of the thorax, for a total of three pairs. The wings themselves are thought by many to be highly modified (tracheal) gills. And there is no doubt that the tracheal gills of the mayfly nymph in many species look like wings. By comparing a well-developed pair of gill blades in the naiads and a reduced pair of hind wings on the adults, it is not hard to imagine that the mayfly gills (tergaliae) and insect wings have a common origin, and newer research also supports this. The tergaliae are not found in any other order of insects, and they have evolved in different directions with time. In some nymphs/naiads the most anterior pair has become sclerotized and works as a gill cover for the rest of the gills. Others can form a large sucker, be used for swimming or modified into other shapes. But it doesn't have to mean that these structures were originally gills. It could also mean that the tergaliae evolved from the same structures which gave rise to the wings, and that flying insects evolved from a wingless terrestrial species with pairs of plates on its body segments: three on the thorax and nine on the abdomen (mayfly nymphs with nine pairs of tergaliae on the abdomen exist, but so far no living or extinct insects with plates on the last two segments have been found). If these were primary gills, it would be a mystery why they should have waited so long to be modified when we see the different modifications in modern mayfly nymphs.

Theories

When the first forests arose on Earth, new niches for terrestrial animals were created. Spore-feeders and others who depended on plants and/or the animals living around them would have to adapt too to make use of them. In a world with no flying animals, it would probably just be a matter of time before some arthropods who were living in the trees evolved paired structures with muscle attachments from their exoskeleton and used them for gliding, one pair on each segment. Further evolution in this direction would give bigger gliding structures on their thorax and gradually smaller ones on their abdomen. Their bodies would have become stiffer while thysanurans, which didn't evolve flight, kept their flexible abdomen. Mayfly nymphs must have adapted to water while they still had the "gliders" on their abdomen intact. So far there is no concrete evidence to support this theory either, but it is one that offers an explanation for the problems of why presumably aquatic animals evolved in the direction they did. Leaping and arboreal insects seems like a good explanation for this evolutionary process for several reasons. Because early winged insects were lacking the sophisticated wing folding mechanism of neopterous insects, they must have lived in the open and not been able to hide or search for food under leaves, in cracks, under rocks and other such confined spaced.

In these old forests there weren't many open places where insects with huge structures on their back could have lived without experiencing huge disadvantages. If insects got their wings on land and not in water, which clearly seems to be the case, the tree canopies would be the most obvious place where such gliding structures could have emerged, in a time when the air was a new territory. The question is if the plates used f[illegible]g evolved from "scratch" or by modifying already ex[illegible]ing anatomical details.

The thorax in Thysanura and Archaeognatha are known to have some structures connected to their trachea which share similarities to the wings of primitive insects. This suggests the origin of both the wings and the spiracles are related. Gliding requires universal body modifications, as seen in present-day vertebrates such as some rodents and marsupials, which have grown wide, flat expansions of skin for this purpose. The flying dragons (genus *Draco*) of Indonesia has modified its ribs into gliders, and even some snakes can glide through the air by spreading their ribs. The main difference is that while vertebrates have an inner skeleton, primitive insects had a flexible and adaptive exoskeleton.

It is clear that there would have been some animals living in the trees, as animals are always taking advantage of all available niches, both for feeding and protection. At the time, the reproductive organs were by far the most nutritious part of the plant, and these early plants show signs of arthropod consumption and adaptations to protect themselves, for example by placing their reproductive organs as high up as possible. But there will always be some species who will be able to cope with that by following the their food source up the trees. Knowing that insects were terrestrial at that time and that some arthropods (like primitive insects) were living in the tree crowns, it seems less likely that they would have developed their wings down on the ground or in the water. In a three dimensional environment such as trees, the ability to glide would increase the insects' chances to survive a fall, as well as saving energy. This trait has repeated itself in modern wingless species such as the gliding ants who are living an arboreal life. When the gliding ability first had originated, gliding and leaping behaviour would be a logical next step, which would eventually be reflected in their anatomical design. The need to navigate through vegetation and to land safely would mean good muscle control over the proto-wings, and further improvements would eventually lead to true (but primitive) wings. While the thorax got the wings, a long abdomen could have served as a stabilizer in flight. It is also worth remembering that some of the earliest flying insects were large predators. This isn't surprising since there weren't yet any other predators hunting in the air: it was therefore a totally new ecological niche. Some of the prey were without a doubt other insects, as insects with proto-wings would have radiated into other species even before the wings were fully evolved. From this point onwards, the arms race could continue: the same predator/prey co-evolution which has existed as long as there have been predators and prey on earth; both the hunters and the hunted were in need of improving and extending their flight skills even further to keep up with the other. Insects that had evolved their proto-wings in a world without flying predators could afford to be exposed openly without risk, but this changed when carnivorous flying insects evolved. It is unknown when they first evolved, but once these predators had emerged they put a strong selection pressure on their victims and themselves. Those of the prey who came up with a good solution about how to fold their wings over their backs in a way that made it possible for them to live in narrow spaces would not only be able to hide from flying predators (and terrestrial predators if they were on the ground) but also to exploit a wide variety of niches that were closed to those who couldn't fold their wings in this way. And today the neopterous insects (those that can fold their wings back over the abdomen) are by far the most dominant group of insects. The water-skimming theory suggests that skimming on the water surface is the origin of insect flight. This theory is based on the fact that the first fossil insects, the Devonian *Rhyniognatha hirsti*, is thought to have possessed wings, even though the insects closest evolutionary ties are with crustaceans, which are aquatic.

Life Cycle

Mayflies

Another primitive trait of the mayflies are the subimago;

no other insects have this winged yet sexually immature stage. A few specialized species have females with no subimago, but retain the subimago stage for males. The reasons the subimago still exists in this order could be that there hasn't been enough selection pressure to get rid of it; it also seems specially adapted to do the transition from water to air. The male genitalia are not fully functional at this point. One reason for this could be that the modification of the abdominal appendages into male copulation organs emerged later than the evolution of flight. This is indicated by the fact that dragonflies have a different copulation organ than other insects. As we know, in mayflies the nymphs and the adults are specialized for two different ways of living; in the water and in the air. The only stage (instar) between these two is the subimago. In more primitive fossil forms, the preadult individuals had not just one instar but numerous ones (while the modern subimago do not eat, older and more primitive species with a subimagos were probably feeding in this phase of life too as the lines between the instars were much more diffuse and gradual than today). Adult form was reached several moults before maturity. They probably didn't have more instars after becoming fully mature.

This way of maturing is how Apterygota do it, which moult even when mature, but not winged insects. Modern mayflies have eliminated all the instars between imago and nymph, except the single instar called subimago, which is still not (at least not in the males) fully sexually mature. The other flying insects with incomplete metamorphosis (Exopterygota) have gone a little further and completed the trend; here all the immature structures of the animal from the last nymphal stage are completed at once in a single final moult. The more advanced insects with larvae and complete metamorphosis (Endopterygota) have gone even further. An interesting theory here is that the pupal stage is actually a strongly modified and extended stage of subimago, but so far it is nothing more than a theory. Interestingly enough there are some insects within the

Exopterygota, thrips and whiteflies (Aleyrodidae), who have evolved pupae-like stages too.

Distant Ancestors

The distant ancestor of flying insects, a species with primitive proto-wings, had a more or less ametabolous life cycle and instars of basically the same type as thysanurans with no defined nymphal, subimago or adult stages as the individual became older. Individuals developed gradually as they were growing and moulting, but there were probably no big changes in between instars. Modern mayfly nymphs do not acquire gills until after their first moult. Before this stage they are so small that there is no need for gills to extract oxygen from the water. This could be a trait from the common ancestor all flyers evolved from. An early terrestrial insect would have no need for paired outgrowths from the body before it started to live in the trees (or in the water, for that matter), so it would not have any. This would also affect the way their offspring looked like in the early instars, resembling earlier ametabolous generations even after they had started to adapt to a new way of living, in a habitat where they actually could have some good use for flaps along their body. Since they matured in the same way as thysanurans with plenty of moultings as they were growing and very little difference between the adults and much younger individuals (unlike modern insects, who are hemimetabolous or holometabolous), there probably wasn't much room for adapting into different niches depending on age and stage. Also, it would have been difficult for an animal already adapted to a niche to make a switch to a new niche later in life based on age or size differences alone when these differences were not significant.

So they had to specialize and focus their whole existence on improving a single lifestyle in a particular niche. The older the species and the single individuals became, the more would they differ from their original form as they adapted to their new lifestyle better than the generations before The final body design was no longer achieved while still inside the egg, but continued to develop for most of the life, causing a bigger difference between the youngest and oldest individuals. Assuming that mature individuals most likely mastered their new element better than did the nymphs who had the same lifestyle, it would appear to be an advantage if the immatures reached adult shape and form as soon as possible. This may explain why they evolved fewer but more intense instars and a stronger focus on the adult body, and the differences between the adults and the first instars were greater, instead of just gradually growing bigger as earlier generations had done. This evolutionary trend explains how they went from ametabolous to hemimetabolous insects. Reaching maturity and a fully grown body became only a part of the development process, gradually also a new anatomy and new abilities only possible in the later stages of life, were included. The anatomy they were born and grew up with had limitations the adults who had learned to fly didn't have. If they couldn't live their early life the way adults did, immature individuals had to adapt to the best way of living and surviving despite their limitations till the moment came when they could leave them behind. This would be a starting point in the evolution where imago and nymphs started to live in different niches, some more clearly defined than others. Also, a final anatomy, size and maturity reached at once with a single final nymphal stage meant less waste of time and energy, and also made a more complex adult body structure. These strategies obviously became very successful with time.

Late Carboniferous and Early Permian insect orders include both several current very long-lived groups (mayflies, Ephemeroptera), dragonflies (Odonata), cockroaches (Blattodea), and Orthoptera (grasshoppers and their relatives) and a number of Paleozoic forms. During this era, some giant dragonfly-like forms – e.g. *Meganeura* and *Meganeuropsis* (Order Protodonata) and *Mazothairos* (Order Palaeodictyoptera) – reached wingspans of 55 to 70 cm (22 to 28 in), making them far larger than any living insect.

Also their nymphs must have had a very impressive size. This gigantism may have been due to higher atmospheric oxygen levels (up to 80% above modern levels during the Carboniferous) that allowed increased respiratory efficiency relative to today. The lack of flying vertebrates could have been another factor. Most extant orders of insects developed during the Permian period that began around 270 million years ago. Many of the early groups became extinct during the Permian-Triassic extinction event, the largest mass extinction in the history of the Earth, around 252 million years ago. The remarkably successful Hymenopterans appeared in the Cretaceous but achieved their diversity more recently, in the Cenozoic. A number of highly successful insect groups — especially the Hymenoptera and Lepidoptera (butterflies), as well as many types of Diptera (flies) and Coleoptera (beetles) — evolved in conjunction with flowering plants, a powerful illustration of co-evolution. Many modern insect genera developed during the Cenozoic; insects from this period on are often found preserved in amber, often in perfect condition. Such specimens are easily compared with modern species. The study of fossilized insects is called paleoentomology.

INSECT PHYSIOLOGY

Insect physiology includes the physiology and biochemistry of insect organ systems. Although diverse, insects are quite similar in overall design, internally and externally. The insect is made up of three main body regions (tagmata), the head, thorax and abdomen. The head comprises six fused segments with compound eyes, ocelli, antennae and mouthparts, which differ according to the insect's particular diet, e.g. grinding, sucking, lapping and chewing. The thorax is made up of three segments the pro, meso and meta thorax, each supporting a pair of legs which may also differ, depending on function, e.g. jumping, digging, swimming and running. Usually the middle and the last segment of the thorax have paired wings. The abdomen generally comprises eleven segments and contains the digestive and reproductive organs (McGavin, 2001). A general overview of the internal structure and physiology of the insect is presented, including digestive, circulatory, respiratory, muscular, endocrine and nervous systems, as well as sensory organs, temperature control, flight and molting.

Digestive System

An insect uses its digestive system to extract nutrients and other substances from the food it consumes. Most of this food is ingested in the form of macromolecules and other complex substances (such as proteins, polysaccharides, fats, and nucleic acids) which must be broken down by catabolic reactions into smaller molecules (i.e. amino acids, simple sugars, etc.) before being used by cells of the body for energy, growth, or reproduction. This break-down process is known as digestion.

The insect's digestive system is a closed system, with one long enclosed coiled tube called the alimentary canal which runs lengthwise through the body. The alimentary canal only allows food to enter the mouth, and then gets processed as it travels toward the anus. The insects alimentary canal has specific sections for grinding and food storage, enzyme production and nutrient absorption (McGavin, 2001; Triplehorn & Johnson, 2005). Sphincters control the food and fluid movement between three regions. The three regions include the foregut (stomatodeum) (27,) the midgut (mesenteron) (13), and the hindgut (proctodeum) (16). In addition to the alimentary canal, insects also have paired salivary glands and salivary reservoirs. These structures usually reside in the thorax (adjacent to the fore-gut). The salivary glands (30) produce saliva, the salivary ducts lead from the glands to the reservoirs and then forward through the head to an opening called the salivarium behind the hypopharynx; which movements of the mouthparts help mix saliva with food in the buccal cavity. Saliva mixes with food which travels through salivary tubes into the mouth, beginning the process of breaking it down. The stomatedeum and proctodeum are invaginations of the epidermis and are lined with cuticle (intima). The mesenteron is not lined with cuticle but with rapidly dividing and therefore constantly replaced, epithelial cells (McGavin, 2001; Triplehorn & Johnson, 2005). The cuticle sheds with every moult along with the exoskeleton (Triplehorn & Johnson, 2005). Food is moved down the gut by muscular contractions called peristalsis (Elzinga, 2004).

1. Stomatodeum (foregut): This region stores, grinds and transports food to the next region (Gullan & Cranston, 2005). Included in this are the buccal cavity, the pharynx, the oesophagus, the crop (stores food), and proventriculus or gizzard (grinds food) (Triplehorn & Johnson, 2005). Salivary secretions from the labial glands dilute the ingested food. In mosquitoes (Diptera), which are blood-feeding insects, anticoagulants and blood thinners are also released here.

2. Mesenteron (midgut): Digestive enzymes in this region are produced and secreted into the lumen and here nutrients are absorbed into the insect's body. Food is enveloped by this part of the gut as it arrives from the foregut by the peritrophic membrane which is a mucopolysaccharide layer secreted from the midgut's epithelial cells (McGavin, 2001). It is thought that this membrane prevents food pathogens from contacting the epithelium and attacking the insects' body (McGavin, 2001). It also acts as a filter allowing small molecules through, but preventing large molecules and particles of food from reaching the midgut cells (Gullan & Cranston, 2005). After the large substances are broken down into smaller ones, digestion and consequent nutrient absorption takes place at the surface the epithelium (McGavin, 2001). Microscopic projections from the mid-gut wall, called microvilli, increase surface area and allow for maximum absorption of nutrients.

3. Proctodeum (hindgut): This is divided into three sections; the anterior is the ileum, the middle portion, the colon, and the wider, posterior section is the rectum (Gullan

& Cranston, 2005). This extends from the pyloric valve which is located between the mid and the hindgut to the anus (Triplehorn & Johnson, 2005). Here absorption of water, salts and other beneficial substances take place before excretion (Gullan & Cranston, 2005). Like other animals, the removal of toxic metabolic waste requires water. However, for very small animals like insects, water conservation is a priority. Because of this, blind-ended ducts called Malpighian tubules come into play (McGavin, 2001). These ducts emerge as evaginations at the anterior end of the hindgut and are the main organs of osmoregulation and excretion (Triplehorn & Johnson, 2005; Gullan & Cranston, 2005). These extract the waste products from the haemolymph, in which all the internal organs are bathed (McGavin, 2001). These tubules continually produce the insect's uric acid, which is transported to the hindgut, where important salts and water are re-absorbed by both the hindgut and rectum. Excrement is then voided as insoluble and non-toxic uric acid granules (McGavin, 2001). Excretion and osmoregulation in insects are not orchestrated by the Malpighian tubules alone, but require a joint function of the ileum and/or rectum (Gullan & Cranston, 2005).

Circulatory System

Insect blood or haemolymph's main function is that of transport and it bathes the insect's body organs. Making up usually less than 25% of an insect's body weight, it transports hormones, nutrients and wastes and has a role in, osmoregulation, temperature control, immunity, storage (water, carbohydrates and fats) and skeletal function. It also plays an essential part in the moulting process (McGavin, 2001; Triplehorn & Johnson, 2005). An additional role of the haemolymph in some orders, can be that of predatory defence. It can contain unpalatable and malodourous chemicals that will act as a deterrent to predators (Gullan & Cranston, 2005). Haemolymph contains molecules, ions and cells (Gullan & Cranston, 2005). Regulating chemical exchanges between tissues, haemolymph is encased in the insect body cavity or haemocoel (Elzinga, 2004; Gullan & Cranston, 2005). It is transported around the body by combined heart (posterior) and aorta (anterior) pulsations which are located dorsally just under the surface of the body (McGavin, 2001; Gullan & Cranston, 2005; Triplehorn & Johnson, 2005). It differs from vertebrate blood in that it doesn't contain any red blood cells and therefore is without high oxygen carrying capacity, and is more similar to lymph found in vertebrates (Elzinga, 2004; Gullan & Cranston, 2005). Body fluids enter through one way valved ostia which are openings situated along the length of the combined aorta and heart organ. Pumping of the haemolymph occurs by waves of peristaltic contraction, originating at the body's posterior end, pumping forwards into the dorsal vessel, out via the aorta and then into the head where it flows out into the haemocoel (Elzinga, 2004; Gullan & Cranston, 2005). The haemolymph is circulated to the appendages unidirectionally with the aid of muscular pumps or accessory pulsatile organs which are usually found at the base of the antennae or wings and sometimes in the legs (Gullan & Cranston, 2005). Pumping rate accelerates due to periods of increased activity (Triplehorn & Johnson, 2005). Movement of haemolymph is particularly important for thermoregulation in orders such as Odonata, Lepidoptera, Hymenoptera and Diptera (Gullan & Cranston, 2005).

Respiratory System

Insect respiration is accomplished without lungs using a system of internal tubes and sacs through which gases either diffuse or are actively pumped, delivering oxygen directly to tissues that need oxygen and eliminate carbon dioxide via their cells (Gullan & Cranston, 2005). Since oxygen is delivered directly, the circulatory system is not used to carry oxygen, and is therefore greatly reduced; it has no closed vessels (i.e., no veins or arteries), consisting of little more than a single, perforated dorsal tube which pulses peristaltically, and in doing so helps circulate the hemolymph inside the body cavity. Air is taken in through spiracles, openings which are positioned laterally in the pleural wall, usually a pair on the anterior margin of the meso and meta thorax, and pairs on each of the eight or less abdominal segments, Numbers of spiracles vary from 1 to 10 pairs (McGavin 2001; Elzinga, 2004; Gullan & Cranston, 2005; Triplehorn & Johnson, 2005). The oxygen passes through the tracheae to the trachioles, and enters the body by the process of diffusion. Carbon dioxide leaves the body by the same process (Triplehorn & Johnson, 2005).

The major tracheae are thickened spirally like a flexible vacuum hose to prevent them from collapsing and often swell into air sacs. Larger insects can augment the flow of air through their tracheal system, with body movement and rhythmic flattening of the tracheal air sacs (Triplehorn & Johnson, 2005). Spiracles are closed and opened by means of valves and can remain partly or completely closed for extended periods in some insects, which minimises water loss (McGavin, 2001; Triplehorn & Johnson, 2005,). There are many different patterns of gas exchange demonstrated by different groups of insects. Gas exchange patterns in insects can range from continuous, diffusive ventilation, to discontinuous gas exchange. Terrestrial and a large proportion of aquatic insects perform gaseous exchange as previously mentioned under an open system. Other smaller numbers of aquatic insects have a closed tracheal system, for example, Odonata, Tricoptera, Ephemeroptera, which have tracheal gills and no functional spiracles. Endoparasitic larvae are without spiracles and also operate under a closed system. Here the tracheae separate peripherally, covering the general body surface which results in a cutaneous form of gaseous exchange. This peripheral tracheal division may also lie within the tracheal gills where gaseous exchange may also take place (Gullan & Cranston, 2005).

Muscular System

Many insects are able to lift twenty times their own

body weight and may jump distances that are many times greater than their own length. This is not because they are strong but because they are so small. Muscle power is proportional to its cross-sectional area. Because the mass (the insect's body), that is moved is in proportion to its volume and the fact that they also have a better leverage system than we humans do, they can jump remarkable distances. (Elzinga, 2004; Triplehorn & Johnson, 2005). The muscular system of insects ranges from a few hundred muscles to a few thousand (Triplehorn & Johnson, 2005). Unlike vertebrates that have both smooth and striated muscles, insects have only striated muscles. Muscle cells are amassed into muscle fibres and then into the functional unit, the muscle (Elzinga, 2004). Muscles are attached to the body wall, with attachment fibres running through the cuticle and to the epicuticle, where they can move different parts of the body including appendages such as wings (Gullan & Cranston, 2005; Triplehorn & Johnson, 2005). The muscle fibre has many cells with a plasma membrane and outer sheath or sarcolemma (Gullan & Cranston, 2005). The sarcolemma is invaginated and can make contact with the tracheole carrying oxygen to the muscle fibre. Arranged in sheets or cylindrically, contractile myofibrils run the length of the muscle fibre. Myofibrils comprising a fine actin filament enclosed between a thick pair of myosin filaments slide past each other instigated by nerve impulses (Gullan & Cranston, 2005). Muscles can be divided into four categories:

1. Visceral: these muscles surround the tubes and ducts and produce peristalsis as demonstrated in the digestive system (Elzinga, 2004).
2. Segmental: causing telescoping of muscle segments required for moulting, increase in body pressure and locomotion in legless larvae (Elzinga, 2004).
3. Appendicular: originating from either the sternum or the tergum and inserted on the coxae these muscles move appendages as one unit. (Elzinga, 2004) These are arranged segmentally and usually

in antagonistic pairs (Triplehorn & Johnson, 2005). Appendage parts of some insects, e.g. the galea and the lacinia of the maxillae, only have flexor muscles. Extension of these structures is by haemolymph pressure and cuticle elasticity (Triplehorn & Johnson, 2005).

4. Flight: Flight muscles are the most specialised category of muscle and are capable of rapid contractions. Nerve impulses are required to initiate muscle contractions and therefore flight. These muscles are also known as neurogenic or synchronous muscles. This is because there is a one to one correspondence between action potentials and muscle contractions. In insects with higher wing stroke frequencies the muscles contract more frequently than at the rate that the nerve impulse reaches them and are known as asynchronous muscles (McGavin, 2001; Gullan & Cranston, 2005).

Flight has allowed the insect to disperse, escape from enemies, environmental harm, and colonise new habitats (McGavin, 2001). One of the insect's key adaptations, the mechanics of flight differ from other flying animals because their wings are not modified appendages (McGavin, 2001; Elzinga, 2004). Fully developed and functional wings occur only in adult insects (Gullan & Cranston, 2005). To fly, gravity and drag (air resistance to movement) has to be overcome (Gullan & Cranston, 2005). Most insects fly by beating their wings and to power their flight they have either direct flight muscles attached to the wings, or an indirect system where there is no muscle to wing connection and instead they are attached to a highly flexible box like thorax (Gullan & Cranston, 2005). Direct flight muscles generate the upward stroke by the contraction of the muscles attached to the base of the wing inside the pivotal point. Outside the pivotal point the downward stroke is generated through contraction of muscles that extend from the sternum to the wing. Indirect flight muscles are attached to the tergum and sternum. Contraction makes the tergum and base of the wing pull down. In turn this movement lever the outer or main part of the wing in strokes upward. Contraction of the second set of muscles, which run from the back to the front of the thorax, powers the downbeat. This deforms the box and lifts the tergum (Gullan & Cranston, 2005).

Endocrine System

Hormones are the chemical substances that are transported in the insect's body fluids (haemolymph) that carry messages away from their point of synthesis to sites that where physiological processes are influenced. These hormones are produced by glandular, neuroglandular and neuronal centres (Gullan & Cranston, 2005). Insects have several organs that produce hormones, controlling reproduction, metamorphosis and moulting (Triplehorn & Johnson, 2005). It has been suggested that a brain hormone is responsible for caste dermination in termites and diapause interruption in some insects (Triplehorn & Johnson, 2005). Four endocrine centers have been identified:

1. Neurosecretory cells in the brain can produce one or more hormones that affect growth, reproduction, homeostasis and metamorphosis (Gullan & Cranston, 2005; Triplehorn & Johnson, 2005).
2. Corpora cardiaca are a pair of neuroglandular bodies that are found behind the brain and on either sides of the aorta. These not only produce their own neurohormones but they store and release other neurohormones including PTTH prothoracicotropic hormone (brain hormone), which stimulates the secretory activity of the prothoracic glands, playing an integral role in moulting.
3. Prothoracic glands are diffuse, paired glands located at the back of the head or in the thorax. These glands secrete an ecdysteroid called ecdysone, or the moulting hormone, which initiates the epidermal moulting process (Gullan & Cranston, 2005). Additionally it plays a role in accessory reproductive

glands in the female, differentiation of ovarioles and in the process of egg production.

4. Corpora allata are small, paired glandular bodies originating from the epithelium located on either side of the foregut. They secrete the juvenile hormone, which regulate reproduction and metamorphosis (Gullan & Cranston, 2005; Triplehorn & Johnson, 2005).

The Nervous System

Insects have a complex nervous system which incorporates a variety of internal physiological information as well as external sensory information (Gullan & Cranston, 2005) Like invertebrates the basic component is the neuron or nerve cell. This is made up of a dendrite with two projections that receive stimuli and an axon, which transmits information to another neuron or organ, like a muscle. As for vertebrates, chemicals (neurotransmitters such as acetylcholine and dopamine) are released at synapses (Gullan & Cranston, 2005). Central nervous system: An insect's sensory, motor and physiological processes are controlled by the central nervous system along with the endocrine system (Gullan & Cranston, 2005). Being the principal division of the nervous system, it consists of a brain, a ventral nerve cord and a subesophageal ganglion. This is connected to the brain by two nerves, extending around each side of the oesophagus.

The brain has three lobes:

- Procerebrum, innervating the compound eyes and the ocelli
- Deutocerebrum, innervating the antennae
- Tritocerebrum, innervating the foregut and the labrum (Gullan & Cranston, 2005; Triplehorn & Johnson, 2005).

The ventral nerve cord extends from the subesophageal ganglion posteriorly (Triplehorn & Johnson, 2005). A layer of connective tissue called the neurolemma covers the brain, ganglia, major peripheral nerves and ventral nerve cords. The head capsule (made up of six fused segments) has six pairs of ganglia. The first three pairs are fused into the brain, while the three following pairs are fused into the subesophageal ganglion. The thoracic segments have one ganglion on each side, which are connected into a pair, one pair per segment. This arrangement is also seen in the abdomen but only in the first eight segments. Many species of insects have reduced numbers of ganglia due to fusion or reduction. Some cockroaches have just six ganglia in the abdomen, whereas the wasp *Vespa crabro* has only two in the thorax and three in the abdomen. And some, like the house fly *Musca domestica*, have all the body ganglia fused into a single large thoracic ganglion. The ganglia of the central nervous system act as the coordinating centres with their own specific autonomy where each may coordinate impulses in specified regions of the insect's body (Triplehorn & Johnson, 2005). Peripheral nervous system: This consists of motor neuron axons that branch out to the muscles from the ganglia of the central nervous system, parts of the sympathetic nervous system and the sensory neurons of the cuticular sense organs that receive chemical, thermal, mechanical or visual stimuli from the insects environment (Gullan & Cranston, 2005). The sympathetic nervous system includes nerves and the ganglia that innervate the gut both posteriorly and anteriorly, some endocrine organs, the spiracles of the tracheal system and the reproductive organs (Gullan & Cranston, 2005).

Sense Organs: Chemical senses include the use of chemoreceptors, related to taste and smell, affecting mating, habitat selection, feeding and parasite-host relationships. Taste is usually located on the mouthparts of the insect but in some insects, such as bees, wasps and ants, taste organs can also be found on the antennae. Taste organs can also be found on the tarsi of moths, butterflies and flies. Olfactory sensilla enable insects to smell and are usually found in the antennae (McGavin, 2001).

Chemoreceptor sensitivity related to smell in some substances, is very high and some insects can detect particular odours that are at low concentrations miles from their original source (Triplehorn & Johnson, 2005). Mechanical senses provide the insect with information that may direct orientation, general movement, flight from enemies, reproduction and feeding and are elicited from the sense organs that are sensitive to mechanical stimuli such as pressure, touch and vibration (Triplehorn & Johnson, 2005). Hairs (setae) on the cuticle are responsible for this as they are sensitive to vibration touch and sound (McGavin, 2001). Hearing structures or tympanal organs are located on different body parts such as, wings, abdomen, legs and antennae. These can respond to various frequencies ranging from 100 to 240 kHz depending on insect species (Triplehorn & Johnson, 2005). Many of the joints of the insect have tactile setae that register movement. Hair beds and groups of small hair like sensilla, determine proprioreception or information about the position of a limb, and are found on the cuticle at the joints of segments and legs.

Pressure on the body wall or strain gauges are detected by the campiniform sensilla and internal stretch receptors sense muscle distension and digestive system stretching (McGavin 2001; Triplehorn & Johnson, 2005,). The compound eye and the ocelli supply insect vision. The compound eye consists of individual light receptive units called ommatidia. Some ants may have only one or two, however dragonflies may have over 10,000. The more ommatidia the greater the visual acuity. These units have a clear lens system and light sensitive retina cells. By day, the image flying insects receive is made up of a mosaic of specks of differing light intensity from all the different ommatidia. At night or dusk, visual acuity is sacrificed for light sensitivity (McGavin, 2001). The ocelli are unable to form focussed images but are sensitive mainly, to differences in light intensity (Triplehorn & Johnson, 2005). Colour vision occurs in all orders of insects. Generally insects see better at the blue end of the spectrum than at the red end. In some orders sensitivity ranges can include ultraviolet (McGavin, 2001). A number of insects have temperature and humidity sensors (McGavin, 2001) and insects being small, cool more quickly than larger animals. Insects are generally considered cold-blooded or ectothermic, their body temperature rising and falling with the environment. However, flying insects raise their body temperature through the action of flight, above environmental temperatures (Elzinga, 2004; Triplehorn & Johnson, 2005,). The body temperature of butterflies and grasshoppers in flight may be 5C or 10C above environmental temperature, however moths and bumblebees, insulated by scales and hair, during flight, may raise flight muscle temperature 20–30C above the environment temperature.

Most flying insects have to maintain their flight muscles above a certain temperature to gain power enough to fly. Shivering, or vibrating the wing muscles allow larger insects to actively increase the temperature of their flight muscles, enabling flight (Triplehorn & Johnson, 2005). Until very recently, no one had ever documented the presence of nociceptors (the cells that detect and transmit sensations of pain) in insects, though recent findings of nociception in larval fruit flies challenges this and raises the possibility that some insects may be capable of feeling pain.

Reproductive System

Most insects have a high reproductive rate. With a short generation time, they evolve faster and can adjust to environmental changes more rapidly than other slower breeding animals (McGavin, 2001). Although there are many forms of reproductive organs in insects, there remains a basic design and function for each reproductive part. These individual parts may vary in shape (gonads), position (accessory gland attachment), and number (testicular and ovarian glands), with different insect groups (Gullan & Cranston, 2005).

Female Reproductive System

The female insect's main reproductive function is to

produce eggs, including the egg's protective coating, and to store the male spermatozoa until egg fertilization is ready. The female reproductive organs include, paired ovaries which empty their eggs (oocytes) via the calyces into lateral oviducts, joining to form the common oviduct. The opening (gonopore) of the common oviduct is concealed in a cavity called the genital chamber and this serves as a copulatory pouch (bursa copulatrix) when mating (Gullan & Cranston, 2005). The external opening to this is the vulva. Often in insects the vulva is narrow and the genital chamber becomes pouch or tube like and is called the vagina. Related to the vagina is a saclike structure, the spermatheca, where spermatozoa are stored ready for egg fertilization. A secretory gland (Gullan & Cranston, 2005; Triplehorn & Johnson, 2005) nourishes the contained spermatozoa in the vagina. Egg development is mostly completed by the insect's adult stage and is controlled by hormones that control the initial stages of oogenesis and yolk deposition (Gullan & Cranston, 2005). Most insects are oviviparous, where the young hatch after the eggs have been laid (Triplehorn & Johnson, 2005). Insect sexual reproduction starts with sperm entry that stimulates oogenesis, meiosis occurs and the egg moves down the genital tract. Accessory glands of the female secrete an adhesive substance to attach eggs to an object and they also supply material that provides the eggs with a protective coating. Oviposition takes place via the female ovipositor (Elzinga, 2004; Triplehorn & Johnson, 2005).

Male Reproductive System

The male's main reproductive function is to produce and store spermatozoa and provide transport to the reproductive tract of the female (Gullan & Cranston, 2005). Sperm development is usually completed by the time the insect reaches adulthood (Triplehorn & Johnson, 2005). The male has two testes, which contain follicles in which the spermatozoa are produced. These open separately into the sperm duct or vas deferens and this stores the sperm

(Gullan & Cranston, 2005). The vas deferentia then unite posteriorally to form a central ejaculatory duct, this opens to the outside on an aedeagus or a penis (Triplehorn & Johnson, 2005). Accessory glands secrete fluids that comprise the spermatophore. This becomes a package that surrounds and carries the spermatozoa, forming a sperm-containing capsule (Gullan & Cranston, 2005; Triplehorn & Johnson, 2005,). Sexual and asexual reproduction Most insects reproduce via sexual reproduction, i.e. the egg is produced by the female, fertilised by the male and oviposited by the female. Eggs are usually deposited in a precise microhabitat on or near the required food (Elzinga, 2004). However, some adult females can reproduce without male input. This is known as parthenogenesis and in the most common type of parthenogenesis the offspring are essentially identical to the mother. This is most often seen in aphids and scale insects (Elzinga, 2004).

Metamorphosis and Insect's Life Cycle

An insect's life-cycle can be divided into three types:

- Ametabolous, no metamorphosis, these insects are primitively wingless where the only difference between adult and nymph is size, e.g. Order: Thysanura (Silverfish) (Triplehorn & Johnson, 2005).
- Hemimetabolous, or incomplete metamorphosis. The terrestrial young are called nymphs and aquatic young are called naiads. Insect young are usually similar to the adult. Wings appear as buds on the nymphs or early instars. When the last moult is completed the wings expand to the full adult size, e.g. Order: Odonata (Dragonflies).
- Holometabolus, or complete metamorphosis. These insects have a different form in their immature and adult stages, have different behaviours and live in different habitats. The immature form is called larvae and remains similar in form but increases in size. They usually have chewing

mouthparts even if the adult form mouth parts suck. At the last larval instar phase the insect forms into a pupa, it doesn't feed and is inactive, and here wing development is initiated, and the adult emerges e.g. Order: Lepidoptera (Butterflies and Moths), (Triplehorn & Johnson, 2005).

Molting

As an insect grows it needs to replace the rigid exoskeleton regularly (McGavin 2001; Triplehorn & Johnson, 2005,). Molting may occur up to three or four times or, in some insects, fifty times or more during its life (McGavin, 2001). A complex process controlled by hormones, it includes the cuticle of the body wall, the cuticular lining of the tracheae, foregut, hindgut and endoskeletal structures (McGavin 2001; Triplehorn & Johnson, 2005). The stages of molting:

1. Apolysis—molting hormones are released into the haemolymph and the old cuticle separates from the underlying epidermal cells. The epidermis increases in size due to mitosis and then the new cuticle is produced. Enzymes secreted by the epidermal cells digest the old endocuticle, not affecting the old sclerotised exocuticle.
2. Ecdysis—this begins with the splitting of the old cuticle, usually starting in the midline of the thorax's dorsal side. The rupturing force is mostly from haemolymph pressure that has been forced into thorax by abdominal muscle contractions caused by the insect swallowing air or water. After this the insect wriggles out of the old cuticle.
3. Sclerotinization—after emergence the new cuticle is soft and this a particularly vulnerable time for the insect as its hard protective coating is missing. After an hour or two the exocuticle hardens and darkens. The wings expand by the force of haemolymph into the wing veins (McGavin, 2001; Triplehorn & Johnson, 2005).

RESPIRATORY SYSTEM OF INSECTS

The respiratory system of insects is the means by which insects obtain oxygen from the air. Air enters the respiratory systems of most insects through a series of external openings called spiracles. These external openings, which act as muscular valves in some insects, lead to the internal respiratory system, a densely networked array of tubes called trachea. The scientific tracheal system within an individual is composed of interconnecting transverse and longitudinal tracheae which maintain equivalent pressure throughout the system. These tracheae branch repeatedly, eventually forming tracheoles, which are blind-ended, water-filled compartments only one micrometer in diameter. It is at this level of the tracheoles that oxygen is delivered to the cells for respiration. The trachea are water-filled due to the permeable membrane of the surrounding tissues. During locomotion, the water level retracts due the increase in concentration of lactic acid in the muscle cells. This lowers the water potential and the water is drawn back into the cells via osmosis and air is brought closer to the muscle cells. The diffusion pathway is then reduced and gases can be transferred more easily.

The Structure of the Spiracle

Insects have spiracles on their exoskeletons to allow air to enter the trachea. In insects, the tracheal tubes primarily deliver oxygen directly into the animals' tissues. The spiracles can be opened and closed in an efficient manner to reduce water loss. This is done by contracting closer muscles surrounding the spiracle. In order to open, the muscle relaxes. The closer muscle is controlled by the central nervous system but can also react to localized chemical stimuli. Several aquatic insects have similar or alternative closing methods to prevent water from entering the trachea. Spiracles may also be surrounded by hairs to minimize bulk air movement around the opening, and thus minimize water loss.

Development of Knowledge

Insects were once believed to exchange gases with the environment continuously by the simple diffusion of gases into the tracheal system. More recently, however, large variation in insect ventilatory patterns have been documented and insect respiration appears to be highly variable. Some small insects do demonstrate continuous respiration and may lack muscular control of the spiracles. Others, however, utilize muscular contraction of the abdomen along with coordinated spiracle contraction and relaxation to generate cyclical gas exchange patterns and to reduce water loss into the atmosphere. The most extreme form of these patterns is termed discontinuous gas exchange cycles (DGC).

EUSOCIALITY

Eusociality is a term used for the highest level of social organization in a hierarchical classification. The lower levels of social organization, presociality, were classified using different terms, including presocial, subsocial, semisocial, parasocial and quasisocial. In analogy with some human societies, groups of specialized individuals are sometimes called castes.

Examples

The most familiar examples are social insects such as ants, bees, and wasps (order Hymenoptera), as well as termites (order Isoptera), all with reproductive queens and more or less sterile workers and/or soldiers. *Austroplatypus incompertus*, a species of weevil native to Australia, is the first beetle to be recognized as a eusocial insect. The only mammalian examples are the naked mole rat and the Damaraland mole rat. Eusociality with biologically sterile individuals represents the most extreme form of kin selection. The analysis of eusociality played a key role in the development of theories in sociobiology. The phenomenon of reproductive specialization is found in various organisms. It generally involves the production of sterile members of the species, which carry out specialized tasks, effectively caring for the reproductive members. It can manifest in the appearance of individuals within a group whose behaviour (and sometimes anatomy) is modified for group defense, including self-sacrificing ("altruism").

History

The term "eusocial" was introduced in 1966 by Suzanne Batra and given a more definitive meaning by E. O. Wilson. It was originally defined to include those organisms (originally, only invertebrates) that had certain features:

1. Reproductive division of labor (with or without sterile castes)
2. Overlapping generations
3. Cooperative care of young

Definition Debates

Subsequent to Wilson's original definition, other authors have sought to expand or narrow the definition of eusociality, focusing on the nature and degree of the division of labor, which was not originally specified. A narrower definition specifies the requirement for irreversibly distinct behavioural groups or castes (with respect to sterility and/or other features), and such a definition excludes all social vertebrates (including mole rats), none of which have irreversible castes. A broader definition allows for any temporary division of labor or non-random distribution of reproductive success to constitute eusociality, and some have accordingly argued that even humans may be considered eusocial. Others believe that the hierarchical classification may not serve much purpose.

Theories of Social Evolution

In spite of the obvious advantages of common foraging and defense, eusocial animals have appeared paradoxical to many theorists of the field of evolution, including W.D. Hamilton: if adaptive evolution unfolds by differential survival of individuals, how can individuals incapable of passing on their genes possibly evolve and persist? Since they do not breed, their fitness should be zero and any genes causing this condition should be eliminated from the population immediately. In *Origin of Species*, Darwin called this behaviour the "one special difficulty, which at first appeared to me insuperable, and actually fatal to my theory." Darwin anticipated that a possible resolution to the paradox might lie in the close family relationship, but specific theories (e.g. kin selection or inclusive fitness) had to wait for the discovery of the mechanisms for genetic inheritance. Early ideas on eusociality included suggestions that trophallaxis or food sharing was a basis for sociality. According to inclusive fitness theory, eusociality may be easier for species like ants to evolve, due to their haplodiploidy, which facilitates the operation of kin selection. Sisters are more related to each other than to their offspring.

This mechanism of sex determination gives rise to what W. D. Hamilton first termed "supersisters" who share 75 percent of their genes on average. Sterile workers are more closely related to their supersisters than to any offspring they might have, if they were to breed themselves. From the "selfish gene's" point-of-view, it is advantageous to raise more sisters. Even though workers often do not reproduce, they are potentially passing on more of their genes by caring for sisters than they would by having their own offspring (each of which would only have 50% of their genes). This unusual situation where females may have greater fitness when they help rear siblings rather than producing offspring is often invoked to explain the multiple independent evolutions of eusociality (arising some 11 separate times) within the haplodiploid group Hymenoptera — ants, bees and wasps. Reeve and Holldobler's version of superorganism theory further elaborates this model by considering competition and co-operation between groups as well as within groups. In this case, an individual's inclusive fitness varies depending on how much it invests in within-group competition (e.g. hoarding a private food cache) versus between-group competition (e.g. contributing to common foraging); and on its relatedness to the other group members. In a hymenopteran colony with one breeder (queen) and many workers as described above, the evolutionarily stable state is for each individual to invest entirely in helping the group, leading to a perfect "superorganism", which implies the stability of eusociality in this case.

This agrees with Hamilton's model. This is implied even without considering between-group interactions. However, they further show that any group of relatives may show high "superorganismness", provided that there are many groups competing for the same resources. This may favour eusociality, or a degree of eusociality in non-hymenopterans. Indeed, a non-zero level of intra-group co-operation is predicted, even if the group members are entirely unrelated, as long as there is competition between groups. Theories of parental manipulation point out that the transition from solitary to eusocial appears to involve intermediate stages where dominance interactions are required to suppress the reproductive tendencies of group members; that is, females are *manipulated* into acting as workers, even if it is against their own self-interest. This model does not require that individuals be highly related, though high relatedness will reduce expected levels of resistance to manipulation. However, now many eusocial species have been discovered that are not haplodiploid (in addition to termites, such as a species of platypodid ambrosia beetles, several independent lines of *Synalpheus* sponge-dwelling shrimp and bathyergid mole rats). Conversely the solitary bees are (as all bees) haplodiploid yet are not eusocial. The association between haplodiploidy and eusociality is below statistical significance, all of which suggests that haplodiploidy alone is neither necessary or sufficient for eusociality to emerge.

Other Examples

Recently, some species of gall-making aphids (Order Hemiptera) and thrips (Order Thysanoptera) were found to be eusocial, with many separate origins of the state. These species have extremely high relatedness among individuals due to their partially asexual mode of reproduction (sterile soldier castes being of the same clone as the reproducing female), but the gall-inhabiting behaviour gives these species a defensible resource that sets them apart from related species with similar genetics. In these groups, therefore, high relatedness alone does not lead to the evolution of social behaviour, but requires that groups occur in a restricted, shared area. Similarly, eusociality has arisen among some crustaceans and other arthropods. On some tropical reefs several species of minute *Synalpheus* pistol shrimp, that depend on certain sponges for the survival of their colony, live eusocially with a single breeding female and a preponderance of male defenders, armed with enlarged snapping claws. Again, there is a single shared domicile for the colony members, and the non-breeding members act to defend it. A single species of beetle, *Austroplatypus incompertus*, is known to be eusocial.

DEFENSE IN INSECTS

Insects have a wide variety of predators, including birds, reptiles, amphibians, mammals, carnivorous plants, and other arthropods. The great majority (80–99.99%) of individuals born do not survive to reproductive age, with perhaps 50% of this mortality rate attributed to predation. In order to deal with this ongoing escapist battle, insects have evolved a wide range of defense mechanisms. The only restraint on these adaptations is that their cost, in terms of time and energy, does not exceed the benefit that they provide to the organism in question. The further that a feature tips the balance towards beneficial, the more likely that selection will act upon the trait, passing it down to further generations. The opposite also holds true; defenses that are too costly will have a slim to zero chance of being passed down. Examples of defenses that have withstood the test of time include hiding, escape by flight or running, and firmly holding ground to fight as well as producing chemicals and social structures that help prevent predation. One of the best known modern examples of the role that evolution has played in insect defenses is the link between melanism and the peppered moth (*Biston betularia*). Peppered moth evolution over the past two centuries in England has taken place, with darker morphs becoming more prevalent over lighter morphs so as to reduce the risk of predation. However, its underlying mechanism is still up to debate.

Hiding

Walking sticks (order Phasmatodea), many katydid species (family Tettigoniidae), and moths (order Lepidoptera) are just a few of the plethora of insects that have evolved specialized cryptic morphology. This adaptation allows them to effectively hide within their environment because of a resemblance to the general background or an inedible object. When an insect looks like an inedible or inconsequential object in the environment that is of no interest to a predator, such as leaves and twigs, it is said to display mimensis, a form of crypsis. Insects may also take on different types of camouflage, another type of cypsis. These include resembling a uniformly colored background as well as being light below and dark above, or coundershaded. Additionally, camouflage is effective when it results in patterns or unique morphologies that disrupt outlines so as to better merge the individual into the background.

Cost and Benefit Perspective

Butterflies (order Lepidoptera) are a good example of the balancing act between the costs and benefits associated with defense. In order to take off, butterflies must have a thorax temperature of 36–40 °C (97–104 °F). This energy is derived both internally through muscles and externally through picking up solar radiation through the body or wings. When looked at in this light, cryptic coloration to escape from predators, markings to attract conspecifics or warn predators (aposematism), and the absence of color to absorb adequate solar radiation, all play key roles in survival. Only when these three affairs are in balance does the butterfly maximize its fitness.

Mimicry

Mimicry is a unique form of defense and describes when a species resembles another recognized by natural enemies, giving it protection against predators. The superficial resemblance among mimics does not necessarily denote common ancestry. Mimicry works if and only if predators are able to learn from eating distasteful species. It is a three part system that involves a model species, a mimic of that species, and a predatory observer that acts as a selective agent. If learning is to be successful, then all models, mimics, and predators must co-exist, a notion feasible within the context of geographic sympatry. Mimicry is divided into two parts, Batesian mimicry and Müllerian mimicry.

Batesian Mimicry

In Batesian mimicry, an aposematic inedible model has an edible mimic. Automimics are individuals that, due to environmental conditions, lack the distasteful or harmful chemicals of conspecifics, but are still indirectly protected through their visibly identical relatives. An example can be found in the Plain Tiger (*Danaus chrysippus*), a non-edible butterfly, which is mimicked by multiple species, the most similar being the female Danaid Eggfly (*Hypolimnas misippus*).

Müllerian Mimicry

In Müllerian mimicry, a group of species benefit from each others' existence because they all are warningly colored in the same manner and are distasteful. The best examples of this phenomenon can be found within the Heliconius butterfly genus. Like in Batesian mimicry, the mimics are not necessarily related, although they obviously are in Heliconius.

Behavioural Responses

Behavioural responses to escape predation include burrowing into substrate and being active only through a restricted period of the day. Furthermore, insects may feign death, a response termed thanatosis. Beetles, particularly weevils, are especially fond of this sort of acting. Bright colors may also be flashed underneath cryptic ones. A startle display occurs when prey takes advantage of these markings after being discovered by a predator. The striking color pattern, which often includes eyespots, is intended to evoke prompt enemy retreat. Better formed eyespots seem to result in better deterrence.

Mechanical Defenses

Insects have had millions of years to evolve a variety of mechanical defenses. Perhaps the most obvious is the cuticle. Although its main role lies in support and muscle attachment, when extensively hardened by the cross-linking of proteins and chitin, or sclerotized, the cuticle acts as a first line of defense. Additional physical defenses include modified mandibles, horns, and spines on the tibia and femur. When these spines take on a main predatory role, they are termed raptorial. Some insects uniquely create retreats that appear uninteresting or inedible to predators. This is the case in caddisfly larvae (order *Trichoptera*) which encase their abdomen with a mixture of natural materials like leaves, twigs, and stones.

Autotomy

Autotomy, or the shedding of appendages, is also used to distract predators, giving the prey a chance to escape. This highly costly mechanism is regularly practiced within stick insects (order Phasmatodea) where the cost is accentuated by the possibility that legs can be lost 20% of the time during molting. Harvestmen (order Opiliones) also use autotomy as a first line of defense against predators.

Chemical Defenses

Unlike pheromones, allomones harm the receiver at

the benefit of the producer. This grouping encompasses the chemical arsenal that numerous insects employ. Insects with chemical weaponry usually make their presence known through aposematism. Aposematism is utilized by non-palatable species as a warning to predators that they represent a toxic danger. Additionally, these insects tend to be relatively large, long-lived, active, and frequently aggregate. Indeed, longer-lived insects are more likely to be chemically defended than short lived ones, as longevity increases apparency. Throughout the arthropod, and insect, realm, however, chemical defenses are quite unevenly distributed. There is great variation in the presence and absence of chemical arms among orders and families to even within families. Moreover, there is diversity among insects as to whether the defensive compounds are obtained intrinsically or extrinsically. Many compounds are derived from the main food source of insect larvae, and occasionally adults, feed, whereas other insects are able to synthesize their own toxins. In reflex bleeding, insects dispel their blood, hemolymph, or a mixture of exocrine secretions and blood as a defensive maneuver. As previously mentioned, the discharged blood may contain toxins produced within the insect source or externally from plants that the insect consumed. Reflexive bleeding occurs in specific parts of the body; for example, the beetle families Coccinellidae (ladybugs) and Meloidae bleed from the knee joints.

Classification

Gullan and Cranston have divided chemical defenses into two classes. Class I chemicals irritate, injure, poison, or drug individual predators. They can be further separated into immediate or delayed substances, depending on the amount of time it takes to feel their effects. Immediate substances are encountered topographically when a predator handles the insect while delayed chemicals, which are generally contained within the insect's tissues, induce vomiting and blistering. Class I chemicals include bufadienolides, cantharidin, cyanides, cardenolides, and alkaloids, all of which have greater effects on vertebrates than on other arthropods. In fact, the most frequently encountered defensive compounds in insects are alkaloids. In contrast to Class I chemicals, Class II chemicals are essentially harmless. They stimulate scent and taste receptors so as to discourage feeding. They tend to have low molecular weight and are volatile and reactive, including acids, aldehydes, aromatic ketones, quinones, and terpenes. Furthermore, they may be aposematic, indicating through odors the presence of chemical defenses. The two different classes are not mutually exclusive, and insects may use combinations of the two.

Pasteels, Grégoire, and Rowell-Rahier also grouped chemical defenses, albeit in a different manner. Weaponry is partitioned into chemical compounds that are truly poisonous, those that restrict movement, and those that repel predators. True poisons, essentially Class I compounds, interfere with specific physiological processes or act at certain sites. Repellents are similar to those classified under Class II as they irritate the chemical sensitivity of predators. Impairment of movement and sense organs is achieved through sticky, slimy, or entangling secretions that act mechanically rather than chemically. This last grouping of chemicals has both Class I and Class II properties. As with Class I and Class II compounds, these three categories are not mutually exclusive, as some chemicals can have multiple effects.

Examples

Assassin Bugs

Some insects inject their venom directly into their offenders, as is the case with the over 5,000 species in the family Reduviidae, the assassin bugs and thread-legged bugs. Interestingly, the assassin bug species Platymeris rhadamanthus is capable of spitting venom up to 30 cm. The saliva of this insect, loaded with a mixture of at least six proteins including excess amounts of protease, hyaluronidase, and phospholipase causes intense local pain, vasodilation, and edema.

Cockroaches

Many cockroach species (order Blattodea) have mucous-like adhesive secretions on their posterior. Although not as affective against vertebrates, these secretions foul the mouths of invertebrate predators, increasing the chances of the cockroach escaping.

Termites

The majority of termite soldiers secrete a rubberlike and sticky chemical concoction that serves to entangle enemies and is usually coupled with specialized mandibles. In nasute species of termites (contained within the subfamily Nasutitermitinae), the mandibles have receded. This makes way for an elongated, syringic nasus capable of squirting liquid glue. When this substance is released from the frontal gland reservoir and dries, it becomes sticky and is capable of immobilizing attackers. It is highly effective against other arthropods, including spiders, ants, and centipedes. Among termite species in the Apicotermitinae that are soldierless or where soldiers are rare, mouth secretions are commonly replaced by abdominal dehiscence. These termites contract their abdominal muscles, resulting in the fracturing of the abdominal wall and the expulsion of gut contents. Because abdominal dehiscence is quite effective at killing ants, the noxious chemical substance released is likely contained within the termite itself.

Ants

Venom is the defense of choice for many ants (family Formicidae). It is injected from an ovipositor that has been evolutionarily modified into a stinging apparatus. These ants release a complex venom mixture that can include histamine. Within the Formicinae subfamily, the stinger has been lost and instead the poison gland forcibly ejects the fluid of choice, formic acid. Some carpenter ants (genus *Camponotus*) also have mandibular glands that extend throughout their bodies. When these are mechanically irritated, the ant forcibly commits suicide and by exploding, spilling out a sticky, entangling substance. The subfamily Dolichoderinae, which also does not possess a stinger, has a different type of defense. The anal gland secretions of this group rapidly polymerize in air and serve to immobilize predators.

Collective Defenses in Social Insects

Many chemically defended insect species take advantage of clustering over solitary confinement. Among some insect larvae in the orders Coleoptera and Hymenoptera, cycloalexy is adopted. Either the heads or ends of the abdomen, depending on where noxious compounds are secreted, make up the circumference of a circle. The remaining larvae lie inside this defensive ring where the defenders repel predators through threatening attitudes, regurgitation, and biting. Termites (order Isoptera), like eusocial ants, wasps, and bees, rely on a caste system to protect their nests. The evolution of fortress defense is closely linked to the specialization of soldier mandibles. Soldiers can have biting-crushing, biting-cutting, cutting, symmetrical snapping, and asymmetrical snapping mandibles. These mandibles may be paired with frontal gland secretion, although snapping soldiers rarely utilize chemical defenses. Termites take advantage of their modified mandibles in phragmosis, which is the blocking of the nest with any part of the body; in this case of termites, nest entrances are blocked by the heads of soldiers.

Alarm pheromones warn members of a species of approaching danger. Because of their altruistic nature, they follow the rules of kin selection. They can illicit both aggregational and dispersive responses in social insects depending on the alarm caller's location relative to the nest. Closer to the nest, it causes social insects to aggregate and may subsequently produce an attack against the threat. In nonsocial insects, these compounds typically stimulate dispersal regardless of location. Chemical alarm systems are best developed in aphids and treehoppers (family Membracidae) among the nonsocial groups. Alarm pheromones take on a variety of compositions, ranging from terpenoids in aphids and termites to acetates, an alcohol, and a ketone in honeybees to formic acid and terpenoids in ants.

Immunity

Insects, like nearly every other organism on Earth, are subject to infectious diseases caused by viruses, bacteria, fungi, protozoa, and nematodes. These encounters can either kill the insect or greatly weaken it. Insects protect themselves against these detrimental microorganisms in two ways. Firstly, the body-enveloping chitin cuticle, in conjunction with the tracheal system and the gut lining, serve as major physical barriers to entry. Secondly, hemolymph itself plays a key role in repairing external wounds as well as destroying foreign organisms within the body cavity. Interestingly, insects, along with having passive immunity, also show evidence of acquired immunity.

Role of Phenotypic Plasticity

Phenotypic plasticity is the capacity of a single genotype to exhibit a range of phenotypes in response to variation in the environment. For example, in Nemoria arizonaria caterpillars, the cryptic pattern changes according to season and is triggered by dietary cues. In the spring, the first brood of caterpillars resembles oak catkins, or flowers. By the summer when the catkins have fallen, the caterpillars discretely mimic oak twigs. No intermediate forms are present in this species, although other members of the *Nemoria* genus, such as *N. darwiniata*, do exhibit transitional forms. In social insects such as ants and termites, members of different castes develop different phenotypes. For example, workers are normally smaller with less pronounced mandibles than soldiers. This type of plasticity is more so determined by cues, which tend to be non-harmful stimuli, than by the environment. Phenotypic plasticity is important because it allows an individual to adapt to a changing environment and can ultimately alter their evolutionary path. It not only plays an indirect role in defense as individuals prepare themselves physically to take on the task of avoiding predation through camouflage or developing collective mechanical traits to protect a social hive, but also a direct one. For example, cues elicited from a predator, which may be visual, acoustic, chemical, or vibrational, may cause rapid responses that alter the prey's phenotype in real time.

INHERITED STERILITY IN INSECTS

Inherited sterility in insects is induced by substerilizing doses of ionizing radiation. When partially sterile males mate with wild females, the radiation-induced deleterious effects are inherited by the F1 generation. As a result, egg hatch is reduced and the resulting offspring are both highly sterile and predominately male. Compared with the high radiation required to achieve full sterility in Lepidoptera, the lower dose of radiation used to induce F1 sterility increases the quality and competitiveness of the released insects as measured by improved dispersal after release, increased mating ability, and superior sperm competition.

History of Inherited Sterility

Area-wide integrated pest management programmes using the sterile insect technique (SIT) as a component have been successful against a number of pest flies or Diptera such as the New World screwworm, *Cochliomyia hominivorax*, various species of tephritidae fruit flies and against tsetse flies (Glossinidae). However, most moths or lepidopterans are more resistant to radiation than dipterans, and as a consequence, the higher dose of radiation required to completely sterilize lepidopterans reduces their performance in the field. One approach to circumvent the negative effects associated with the high radio-resistance of Lepidoptera pests has been the use of inherited sterility or F1 sterility, first documented in studies on the codling moth (*Cydia pomonella*). Inherited sterility has also been documented in the Hemiptera order. The silk worm *Bombyx mori* (Lepidoptera: Bombycidae) was the fist insect in which inherited sterility was reported. Then inherited sterility was reported in the greater wax moth *Galleria mellonella* (Lepidoptera: Pyralidae), in the codling moth *Cydia pomonella* (Lepidoptera: Tortricidae), in the large milkweed bug *Oncopeltus fasciatus* (Hemiptera: Lygaeidae), in *Gonocerus acuteangulatus* (Hemiptera: Coreida), in *Rhodnius prolixus* (Hemiptera: Reduviidae), and in the Two-spotted spider mite *Tetranychus urticae* (Acari: Tetranychidae).

Genetic Basis of Inherited Sterility

The mechanisms by which mutations cause lethality in Diptera in the developing zygote are well documented,,. The primary lesion leading to a dominant lethal mutation is a break in the chromosome, in this case, induced by radiation. When a break is induced in a chromosome in mature sperm, it remains in this condition until after the sperm has entered an egg. Following fusion, nuclear divisions begin, and a break in a chromosome can have drastic effects on the viability of the embryo as development proceeds. During early prophase the broken chromosome undergoes normal replication, but during metaphase the broken ends can fuse leading to the formation of a dicentric chromosome and an acentric fragment. The acentric fragment is frequently lost, while the dicentric fragment forms a bridge at anaphase leading to another chromosomal break. This whole process then repeats itself, leading to the accumulation of serious imbalances in the genetic information of the daughter cells. The accumulation of this genetic damage finally leads to the death of the zygote. Diptera, Hymenoptera, and Coleoptera orders can be classed as radiation-sensitive, while Lepidoptera, Homoptera and mites (Acari) orders are radiation-resistant. A major difference between these two groups of Insects is that the former group has a localized centromere (monokinetic), while the latter has a diffuse centromere (holokinetic). However, more recent work suggested that lepidopteran chromosomes are intermediate between holokinetic and monocentric chromosomes. In any case, the centromere difference is believed to play a major, although not exclusive, role in radiation sensitivity. It was suggested, that possible molecular mechanisms responsible for the high radioresistance in Lepidoptera might include an inducible cell recovery system and a DNA repair probes.

Lepidoptera also do not show the classical breakage-fusion-bridge cycle that is a characteristic of dominant lethals induced in Diptera. It appears that lepidopteran chromosomes can tolerate telomere loss without the drastic effects that this has on chromosomes in other orders. Lepidopteran chromosomes possess a localized kinetochore plate to which the spindle microtubules attach during cell division. The kinetochore plates are large and cover a significant portion of the chromosome length, ensuring that more radiation-induced breaks will not lead to the loss of chromosome fragments as is typical in species with monocentric chromosomes. In species with large kinetochore plates, the fragments may persist for a number of mitotic cell divisions, and can even be transmitted through germ cells to the next generation. The plates also reduce the risk of lethality caused by the formation of dicentric chromosomes, acentric fragments,. and other unstable aberrations.

Field Application of Inherited Sterility

The F1 sterile progeny produced in the field enhance the efficacy of released partially sterile males, and improve compatibility with other pest control strategies. For example, the presence of F1 sterile progeny can be used to increase the build-up of natural enemies in the field. In addition, F1 sterile progeny can be used to study the potential host and geographical ranges of exotic lepidopteran pests. Field programmes releasing irradiated moths under an SIT or inherited sterility approach have been in operation since the 1960s. The pink bollworm, *Pectinophora gossypiella* has been successfully contained since 1969 in cotton areas of the San Joaquin Valley in California and is being successfully targeted for eradication from cotton areas in the south-western USA and north-western Mexico. Since the early 1990s, the codling moth has been successfully suppressed in apple and pear production areas in the Okanagan Valley in British Columbia, Canada, and countries such as Argentina, Brazil and South Africa have plans or programmes against this pest. New Zealand eradicated outbreaks of the Australian painted apple moth, *Teia anartoides*. Mexico eradicated outbreaks of the cactus moth, *Cactoblastis cactorum* and the USA contains its advance along the Gulf of Mexico coast. South Africa has a programme to suppress the false codling moth, *Thaumatotibia leucotreta* in citrus orchards Control of most moth pests is hampered by the increased resistance to the most widely used broad spectrum insecticides; hence the potential for expanded implementation of inherited sterility as part of an area-wide integrated approach is considerable.

STERILE INSECT TECHNIQUE

Sterile insect technique is a method of biological control, whereby overwhelming numbers of sterile insects are released. The released insects are normally male as it is the female that causes the damage, usually by laying eggs in the crop, or, in the case of mosquitoes, taking a bloodmeal from humans. The sterile males compete with the wild males for female insects. If a female mates with a sterile male then it will have no offspring, thus reducing the next generation's population. Repeated release of insects can eventually wipe out a population, though it is often more useful to consider controlling the population rather than eradicating it. The technique has successfully been used to eradicate the Screw-worm fly (*Cochliomyia hominivorax*) in areas of North America. There have also been many successes in controlling species of fruit flies, most particularly the Medfly (*Ceratitis capitata*), and the Mexican fruit fly (*Anastrepha ludens*). Insects are mostly sterilized with radiation, which might weaken the newly sterilized insects, if doses are not correctly applied, making them less able to compete with wild males. However, other sterilization techniques are under development which would not affect the insects' ability to compete for a mate. The technique was pioneered in the 1950s by American entomologists Dr. Raymond C. Bushland and Dr. Edward F. Knipling. For their achievement, they jointly received the 1992 World Food Prize.

Development of the Sterile Insect Technique

Raymond Bushland and Edward Knipling first developed the technique to eliminate screwworms preying on warm-blooded animals, especially cattle herds. With larvae that invade open wounds and eat into animal flesh, the flies were capable of killing cattle within 10 days of infection. In the 1950s, screwworms caused annual losses to American meat and dairy supplies that were projected at above $200 million. Screwworm maggots are also known to parasitize human flesh. Since a female screwworm mates only once in her lifetime, this physiological phenomenon has been exploited by biologists in breaking its life cycle. After mating with a sterile male, the screwworm female will not mate again or lay any eggs. The quest of Bushland and Knipling to find an alternative to chemical pesticides in controlling the devastation wrought by these insects began in the late 1930s when both scientists were working at the United States Department of Agriculture Laboratory in Menard, Texas. At that time, the screwworm was decimating livestock herds across the American South. Red meat and dairy supplies were also affected across Mexico, Central America, and South America.

While Bushland initially researched chemical treatment of screwworm-infested wounds in cattle, Knipling developed the theory of autocidal control – breaking the life cycle of the pest itself. Bushland's enthusiasm for Knipling's theory sparked both men to intensify the search for a way to rear large numbers of flies in a "factory" setting, and most importantly, to find an effective way to sterilize flies. Their work in this area was interrupted by World War II, but Drs. Bushland and Knipling resumed their efforts in the early 1950s with their successful tests on the screwworm population of Sanibel Island, Florida. The sterile insect technique worked; near eradication was achieved using X-ray sterilized flies. In 1954, the technique was used to completely eradicate screwworms from the 176-square-mile (460 km) island of Curaçao, off the coast of Venezuela. Screwworms were eliminated in a span of only seven weeks, saving the domestic goat herds that were a source of meat and milk for the island people. During the 1960s and 1970s, SIT was used to control the screwworm population in the United States. The 1980s saw Mexico and Belize eliminate their screwworm problems through the use of SIT, and eradication programmes have progressed through all of Central America, with a biological barrier having been established in Panama to prevent reinfestation from the south. In 1991, Knipling and Bushland's technique halted a serious outbreak in northern Africa.

Similar programmes against the Mediterranean fruit fly in Mexico and California use the same principles. In addition, the technique was used to eradicate the melon fly from Okinawa and has been used in the fight against the tsetse fly in Africa. The technique has been able to suppress insects threatening livestock, fruit, vegetable, and fiber crops. The technique has also been lauded for its many environmentally sound attributes: it uses no chemicals, leaves no residues, and has no effect on non-target species. Proven effective in controlling outbreaks of a wide range of insect pests throughout the world, the technique has been a boon in protecting the agricultural products to feed the world's human population. Both Bushland and Knipling received worldwide recognition for their leadership and scientific achievements, including the World Food Prize. Their research and the resulting Sterile Insect Technique were hailed by former U.S. Secretary of Agriculture Orville Freeman as "the greatest entomological achievement of (the 20th) century."

Sterile Fly for African Trypanosomiasis

Sleeping sickness or the African trypanosomiasis is a parasitic disease in humans. Caused by protozoa of genus *Trypanosoma* and transmitted by the Tsetse fly, the disease is endemic in certain regions of Sub-Saharan Africa, covering about 36 countries and 60 million people. It is estimated that 50,000 - 70,000 people are infected, and about 40,000 die every year. Three major epidemics have occurred in the past hundred years, in 1896 - 1906, 1920, and 1970. Studies of the tsetse fly show that females generally only mate once in their lifetimes and very rarely mate a second time.

Once a female fly has mated, she can then produce continual offspring throughout her short life. The sterile fly is an innovative solution to the problem of the African trypanosomiasis. Specially bred male Tsetse flies are sterilized through irradiation process. These sterilized male flies are then released into areas where sleeping sickness is prevalent, and then mate with the females. Because the male is sterile, and the females mate only once, the population of Tsetse flies in the affected area will drop. Studies have shown that this process has been very effective in preventing sleeping sickness in people who live in the area. Since sleeping sickness is fatal without treatment and infected people can be without symptoms for months, the release of sterile flies into affected areas leads to greater levels of health and economic activity.

Success Stories

- Screwworm fly *Cochliomyia hominivorax* - eradicated from the United States, Mexico, and Libya.
- Mexican fruit fly *Anastrepha ludens* (Loew) eradicated from most of northern Mexico.
- Tsetse fly eradicated from Zanzibar.
- Medfly *Ceratitis capitata* (Wiedemann) - eradicated from northern part of Chile and southern part of Peru and southern part of Mexico.

- Melon fly *Bactrocera cucurbitae* (Coquillett) eradicated from Okinawa, Japan.

Current Targets

- *Anopheles* mosquito - Malaria vector, example *Anopheles arabiensis*.
- Tsetse fly (*Glossina spp*) - sleeping sickness vector.
- Painted Apple Moth (Lepidoptera: Lymantriidae) in Auckland, New Zealand.
- *Aedes* mosquitoes, vectors for filariasis, Dengue and yellow fever.
- Continuing use across the world against various fruit fly species, including Medfly, Caribbean and Mexican fruit fly (Nth, Sth and Central America), Queensland fruit fly in Australia (*Bactrocera tryoni*) and several other *Bactrocera* sp. across Australia, Asia and Oceania.

The IAEA lists 36 different fruit fly SIT facilities across the globe, on all six inhabited continents and provides information on the doses of radiation used in the control of pest insects and mites of agricultural, commercial or quarantine significance. It includes data on both the radiation dose required for the disinfestation of generic commodity groups, fresh and durable, and also the radiation dose used to induce sterility for pest control through the sterile insect technique.

Drawbacks

- As for insecticide treatment, repeated treatment is sometimes required to suppress the population before the use of sterile insects.
- Sex separation could be difficult for some species, though this can be easily performed on Medfly and screwworm, for example.
- Radiation treatment in some case, over doses, affects the health of males, so sterilized insects in such

cases are at a disadvantage when competing for females.

- The technique is species specific, for instance: there are 22 species of Tsetse fly in Africa, and the technique must be implemented separately for each.
- Standard operating procedures of mass rearing and irradiation, do not leave room for mistakes. Since the fifties, when SIT was first used as a means for pest control, several failures have occurred in different places around the world where non-sterilized artificial produced insects were released before the problem was spotted.
- Application to large areas should be long lasting, otherwise migration of wild insects from outside the control area could repopulate.
- The major drawback to this technique is that the cost of producing such a large number of sterile insects is often prohibitive in poorer countries.

Genetic Modification

A method using recombinant DNA technology to create genetically modified insects called RIDL (Release of Insects carrying a Dominant Lethal) is under development by a company called Oxitec. The method works by introducing a repressible "Dominant Lethal" gene into the insects. This gene kills the insects but it can be repressed by an external additive, which allows the insects to be reared in manufacturing facilities. This external additive is commonly administered orally, and so can be an additive to the insect food. The insects can also be given genetic markers, such as fluorescence, that make monitoring the progress of eradication easier. There are potentially several types of RIDL, but the more advanced forms have a female-specific dominant lethal gene. This avoids the need for a separate sex separation step, as the repressor can be withdrawn from the final stage of rearing, leaving only males. These males are then released in large numbers into the affected region. The released males are not sterile, but any female offspring their mates produce will have the dominant lethal gene expressed, and so will die. The number of females in the wild population will therefore decline, causing the overall population to decline. Using RIDL means that the males will not have to be sterilized by radiation before release, making the males more healthy when they need to compete with the wild males for mates. Progress towards applying this technique to mosquitos has been made by researchers at Imperial College London who created the world's first transgenic malaria mosquito. A similar technique is the daughterless carp, a genetically modified organism produced in Australia by the CSIRO in the hope of eradicating the introduced carp from the Murray River system. As of 2005, it was undergoing tests to assess the risks of releasing it into the wild.

Conclusion and Perspectives

Biotechnological approaches based on genetically modified organism (transgenic organisms) are still under development. However, since no legal framework exists to authorize the release of such organisms in the nature, sterilization by irradiation remains the most used technique. A meeting was held at FAO headquarters in Rome, 8 to 12 April 2002 on *"Status and Risk Assessment of the Use of Transgenic Arthropods in Plant Protection"*. The resulting proceedings of the meeting have been used by the North American Plant Protection Organization (NAPPO) to develop NAPPO Regional Standard No. 27 on *"Guidelines for Importation and Confined Field release of Transgenic Arthropods"*, which might provide the basis for the rational development of the use of transgenic arthropods.

SIT programmes will benefit tremendously if genetic methods can be developed that enable only male insects to be reared as has already been done for the medfly. In addition, more appropriate artificial diets for larvae, and hormonal, nutritional, microbiological, and semiochemical treatments for adults, could make major contributions through improved economy and insect quality. Economic benefits of SIT has been demonstrated in various cases. For example, direct benefits of screwworm eradication to the North and Central American livestock industries are estimated to be over $ 1.5 billion/ year, compared with a total investment over half a century of close to $ 1 billion. Mexico protects a fruit and vegetable export market of over $ 3 billion/year through an annual investment of ca. $ 25 million, and medfly-free status has been estimated to have opened markets for Chile's fruit exports of up to $ 500 million. Eradication of tsetse has resulted in major socio-economic benefits for Zanzibar. When implemented on an area-wide basis and with economies of scale in the mass rearing process, the use of SIT for suppression is cost competitive with conventional control, in addition to its environmental benefits.

REFERENCES

Alley N.F., Frakes L.A. (2003). "First known Cretaceous glaciation: Livingston Tillite, South Australia". *Australian Journal of Earth Science* 50: 134–150.

Antonio Arillo, Vicente M. Ortuño (2008). "Did dinosaurs have any relation with dung-beetles? (The origin of coprophagy)". *Journal of Natural History* 42 (19&20): 1405–1408.

Astaurov, B.I., and S.L. Frolova (1935). Artificial mutations in the silkworm (*Bombyx mori* L.). V. Sterility and spermatogenic anomalies in the progeny of irradiated moths concerning some questions of general biological and mutagenic action of X-rays. Biol. J. 4: 861-894 (in Russian).

Batra, S. W.T. (1966). Nests and social behaviour of halictine bees of India (Hymenoptera: Halictidae). — Indian J. Entomol 28 375-393.

Bauer H. (1967). Die kinetische Organization der Lepidopteren-Chromosomen. Chromosoma 22: 101-125.

Beckemeyer, R. J.; M. S. Engel (2008). "A Second Specimen of Permocoleus (Coleoptera) from the Lower Permian Wellington Formation of Noble County, Oklahoma". *Journal of the Kansas Entomological Society* 81 (1): 4–7.

Benisch, Christoph (2010). "Phylogeny of the beetles". *The beetle fauna of Germany*. Kerbtier.

Blum, M.S. (1981). *Chemical Defenses of Arthropods*. New York: Academic Press, Inc.

Burda, H. Honeycutt, R. L, Begall, S., Locker-Grutjen, O & Scharff A. (2000). Are naked and common mole-rats eusocial and if so, why? Behavioural ecology and sociobiology 47(5):293-303 Abstract.

Carpenter, J.E., S. Bloem and F. Marek. (2005). Inherited Sterility in Insects. Chapter 2.4 in Dyck, V.A., J. Hendrichs, and A.S. Robinson, editors, "Sterile Insect Technique. Principles and Practice in Area-Wide Integrated Pest Management". Springer. The Netherlands.

Chin, K.; Gill, B. D. (1996). "Dinosaurs, dung beetles, and conifers; participants in a Cretaceous food web". *PALAIOS* (11): 280–285.

Christopher O'Toole (2002). *Firefly Encyclopedia of Insects and Spiders*. Toronto: Firefly Books.

Cilek JE, Petersen JL, Hallmon C.E. (2004). "Comparative efficacy of IR 3535 and deet as repellents against adult Aedes aegypti and Culex quinquefasciatus.". *J Am Mosq Control Assoc* 20 (3): 299–304.

Collins S.R., Weldon C.W., Banos C., Taylor P.W. (2008). Effects of irradiation dose rate on quality and sterility of Queensland fruit flies, Bactrocera tryoni (Froggatt). Journal of Applied Entomology 132 (5): 398-405.

Collins, D.A.; Brady, J.N.; Curtis, C.F. (1993). "Assessment of the efficacy of Quwenling as a Mosquito repellent". *Phytotherapy Research* 7 (1): 17–20.

D.S. Kent & J.A. Simpson (1992). "Eusociality in the beetle *Austroplatypus incompertus* (Coleoptera: Curculionidae)". *Naturwissenschaften* 79: 86–87.

David Grimaldi & Michael S. Engel (2005). *Evolution of the Insects*. Cambridge University Press.

Duncan, Carl D. (1939). *A Contribution to The Biology of North American Vespine Wasps* (1 ed.). Stanford: Stanford University Press, pp. 24–29.

Dyck, V.A., Hendrichs, J., and Robinson, A.S. (2005). " Sterile Insect Technique: Principles and Practice in Area-wide Integrated Pest Management. " 787 pp. Springer, Dordrecht, The Netherlands.

Elias, S.A. (1996). "Late Pleistocene and Holocene Seasonal Temperatures Reconstructed from Fossil Beetle Assemblages in the Rocky Mountains". *Quaternary Research* 46 (3): 311–318.

Elias, S.A. (2000). "Climatic Tolerances and Zoogeography of the Late Pleistocene Beetle Fauna of Beringia". *Géographie physique et Quaternaire* 54 (2): 143–155.

Elias, S.A. (2000). "Late Pleistocene Climates of Beringia, Based on Analysis of Fossil Beetles". *Quaternary Research* 53 (2): 229–235.

Elzinga, R.J. (2004). *Fundamentals of Entomology* (6th Ed) Pearson / Prentice Hall, New Jersey USA

Frakes L.A., Francis J.E. (1988). "A guide to Phanerozoic cold climates from high latitude ice rafting in the Cretaceous". *Nature* 333 (6173): 547–549.

Gadagkar, Raghavendra (1993). And now... eusocial thrips!. Current Science 64(4): pp. 215-216.

Gillott, C. (1995). *Entomology* (2 ed.). New York: Plenum Press.

Gullan, D.J. and Cranston, P.S. (2005). *The Insects: An Outline of Entomology* Blackwell Publishing Ltd., UK.

Gullan, P. J.; P.S. Cranston (2005). *The Insects: an Outline of Entomology* (3 ed.). Malden, MA: Blackwell Publishing.

H.H. Shorey (1973). "Behavioural responses to insect pheromones". *Annual Review of Entomology* 18: 349–380.

Henneberry, T. J. (1964). Effects of gamma radiation on the fertility of the two-spotted spider mite and its progeny. J. Econ. Entomol. 57:672–674.

J.E. Huheey (1984). "Warning coloration and mimicry". In William J. Bell & Ring T. Cardé. *Chemical Ecology of Insects*. London: Chapman and Hall, pp. 257–297.

J.M. Pasteels, J.C. Grégoire & M. Rowell-Rahier (1983). "The chemical ecology of defense in arthropods" (PDF). *Annual Review of Entomology* 28: 263–289.

James T. Costa & Terrence D. Fitzgerald (2005). Social terminology revisited: Where are we ten years later? Ann. Zool. Fennici 42:559-564.

Joachimski, M.M.; Breisig, S.; Buggisch, W.; Talent, J.A.; Mawson, R..; Gereke, M.; Morrow, J.R.; Day, J. et al. (2009). "Devonian climate and reef evolution: Insights from oxygen isotopes in apatite". *Earth and Planetary Science Letters* 284 (3–4): 599--596.

John R. Meyer (March 8, 2005). "Trichoptera". *ENT 425 - General Entomology*. North Carolina State University.

Kumano Norikuni; Haraguchi Dai; Kohama Tsuguo (2008). Effect of irradiation on mating performance and mating ability in the West Indian sweetpotato weevil, Euscepes postfasciatus. Entomologia Experimentalis et Applicata, Volume 127, Number 3, pp. 229-236(8).

Kumano Norikuni; Kawamura Futoshi; Haraguchi Dai; Kohama Tsuguo (2008). Irradiation does not affect field dispersal ability in the West Indian sweetpotato weevil, Euscepes postfasciatus. Entomologia Experimentalis et Applicata Volume 130 (1) : 63-72.

LaChance, L. E. and Degrugillier, M. (1969). Chromosomal fragments transmitted through three generations in *Oncopeltus* (Hemiptera). Science, 166: 236 237.

Maudlin I. (1976). The inheritance of radiation induced semi-sterility in Rhodnius prolixus. Chromosoma 58 (1976), pp. 285–306.

McGavin, G.C. (2001). *Essential Entomology; An order by order introduction*. Oxford University Press, New York.

Michael S. Engel & David A. Grimaldi (2004). "New light shed on the oldest insect". *Nature* 427 (6975): 627–630.

Nation, James L. (2002). *Insect Physiology and Biochemistry*. Boca Raton, FL: CRC Press.

Oliver Béthoux (2009). "The earliest beetle identified". *Journal of Paleontology* 83 (6): 931–937.

Ostriakova-Varshaver, V.P. (1937). The bee moth, *Galleria mellonella*, as a new object for genetic investigations. II. Cytogenetic analysis of sterility initiated by X-rays in males. [In Russian.] Biologicheskii Zhurnal, 6: 816–836.

Pascal Laurent, Jean-Claude Braekman and Désiré Daloze (2005). *Insect chemical defense*. In Stefan Schulz. "The Chemistry of Pheromones and Other Semiochemicals II". *Topics in Current Chemistry* 240: 167–229.

Ponomarenko, A. G. (1997). "New Beetles of the Family Cupedidae from the Mesozoic of Mongolia. Ommatini, Mesocupedini, Priacmini" (PDF). *Paleontological Journal* 31 (4): 389–399.

Ponomarenko, A. G. (2004). "Beetles (Insecta, Coleoptera) of the Late Permian and Early Triassic". *Paleontological Journal* 38 (Suppl. 2): S185–S196.

Powell, Jerry A. (2009). "Coleoptera". In Resh, Vincent H.; Cardé, Ring T. *Encyclopedia of Insects* (2 (illustrated) ed.). Academic Press, p. 1132.

Resh, Vincent H.; Ring T. Carde (July 1, 2009). *Encyclopedia of Insects* (2 ed.). U. S. A.: Academic Press.

Shcherbakov, D.E. (2008). "On Permian and Triassic Insect Faunas in Relation to Biogeography and the Permian-Triassic Crisis". *Paleontological Journal* 42 (1): 15–31.

Tara Lynne Maginnis (2008). "Autotomy in a stick insect (Insecta: Phasmida): predation versus molting" (PDF). *Florida Entomologist* 91 (1): 126–127.

Taverne, Janice (2001). "Malaria on the Web and the mosquito-repellent properties of basil". *Trends in Parasitology* 17 (6): 299–300.

Tracey, J., W. Daniel, R. I. Wilson, G. Laurent, and S. Benzer (2003). *Painless*, a *Drosophila* gene essential for nociception. Cell 113: 261-273.

Triplehorn, C.A. and Johnson, N.F. (2005). *Borror and DeLong's Introduction to the Study of Insects* (7th Ed). Brooks / Thomson Cole USA.

Whitman, D.W., and L. Blaustein (2009), "Natural enemy-induced plasticity in plants and animals." *Phenotypic Plasticity of Insects: Mechanisms and Consequences.* Ed. D. W. Whitman and T. N. Ananthakrishnan. Enfield, NH: Science, 1-63.

4
Scientific Classification, Subclasses, Orders of Insects and Entomophagy

ARCHAEOGNATHA

Archaeognatha is an order of wingless insects, also known as *jumping bristletails*. They are among the least evolutionarily changed insects, appearing in the Devonian period along with the arachnids. The name Archaeognatha is derived from Greek *Archaeos* meaning "ancient" and *gnatha* meaning "jaw". This refers to the articulation of the mandibles, which has a single condyle, where all higher insects have two. An alternate name, Microcoryphia comes from the Greek *micro* meaning "small" and *coryphia* meaning head. The Order Archaeognatha has previously been combined with the Order Thysanura, or bristletails, both of which groups possess three-pronged tails comprising two cerci and an epiproct. There are approximately 350 species in the two families, distributed worldwide. There are no species at current conservation risk, though the order is one of the most poorly-studied among insects, and therefore it may simply be that no one has yet recognized that any species are at risk.

Description

Archaeognaths are small insects with elongated bodies bent into an arch shape and three long tail-like structures. They have flexible antennae, large compound eyes that meet at the top of the head, and three ocelli. Their mouthparts are partially retractable, with simple chewing mandibles and long maxillary palps. Archaeognatha differ from Thysanura in being able to use their tail to spring up to 30 centimetres (12 in) into the air. They are also unique in possessing small, articulated "styli" on the hind (and sometimes middle) coxae and sternites 2 to 9, which are sometimes considered to be rudimentary appendages, as well as paired eversible membranous vesicles through which they absorb water. Further unusual features are that the abdominal sternites are each composed of three sclerites, and they cement themselves to the substrate before molting. Like Thysanura, the body is covered with scales, with a thin exoskeleton that is susceptible to dehydration.

Biology

Archaeognaths are found in a wide range of habitats, and are unusual among insects in that they can even be found in the Arctic where they live in leaf litter and rock crevices. While most species are found in moist soil, others have adapted to chaparral, and even sandy deserts. They feed primarily on algae, but also lichens, mosses, or decaying organic materials. During courtship, the males spin a thread from their abdomen, attach one end to the substrate, and string packages of sperm (spermatophores) along it. After a series of courtship dances, the female picks up the spermatophores and places them on her ovipositor. The female then lays a batch of around 30 eggs in a suitable crevice. The young resemble the adults, and take up to two years to reach sexual maturity. Unlike most insects, the adults continue to moult after reaching adulthood, and typically mate once at each instar. Archaeognaths may have a total lifespan of up to four years, longer than many larger insects.

COXOPLECTOPTERA

Coxoplectoptera or "chimera wings" is a primitive, extinct order of winged insects containing one family, Mickoleitiidae, described by Staniczek, Bechly & Godunko in 2011. Two adult and more than 20 larval fossils have been scientifically described from Mesozoic outcrops, mainly from the Lower Cretaceous Crato Formation of Brazil (totally at least 30-40 fossil larvae have been found). Coxoplectoptera belong to the stem group of mayflies. Winged adults and the aquatic larvae were predators with raptorial forelegs, which are reminiscent to those of praying mantids. The larvae had a peculiar freshwater shrimp-like habitus.

Etymology

The genus *Mickoleitia* and family Mickoleitiidae was named in honor of German zoologist Gerhard Mickoleit from the University of Tübingen, who was among the first proponents of Willi Hennig's "Phylogenetic Systematics". The scientific name of the order Coxoplectoptera refers to the prolonged coxal segment of the larval and adult legs, and the old scientific name Plectoptera for mayflies (not to be confused with Plecoptera for stoneflies). The common name "chimera wings" was coined in reference to the strange combination of characters in the morphology of the adult animal, which looks like a kind of chimera built from unrelated insects, with their oblique thorax and broad hind wing shape like a dragonfly, their wing venation like a primitive mayfly ancestor, and their raptorial forelegs like a mantis.

History of Discovery

The fossil larvae of the genus *Mickoleitia* are not extraordinarily rare in the limestones of the Crato Formation, so that the local brick workers even have a common Brazilian name for them ("Abacaxi" = pineapple). These larvae were scientifically discovered and first mentioned by Bechly, who also pointed to their strange morphology. Staniczek (2002, 2003) discussed the larvae as well and claimed that they arguably had been a kind of living fossil in the Lower Cretaceous. The German biologist Rainer Willmann described the larvae in a chapter in Martill, Bechly & Loveridge (2007) and erroneously attributed them to the extinct stem group mayfly family Cretereismatidae that he described based on adult specimens from the same locality. During the work for this monograph on the Crato Formation the German palaeoentomologist Günter Bechly and entomologist Arnold H. Staniczek discovered in the fossil collection of the Stuttgart State Museum of Natural History the very adult specimen that later would become the holotype of *Mickoleitia longimanus*. They figured this fossil in Martill, Bechly & Loveridge 2007 as undescribed stem group mayfly and indicated in a brief figure legend the possible relationship to the erratic larvae. The detailed scientific description of Coxoplectoptera and the demonstration of the relationship of fossil adult and larvae was finally published by Staniczek, Bechly & Godunko (2011) in a special issue on Cretaceous insects of the journal "Insect Systematics & Evolution". The authors also determined that two fossil larvae (*Mesogenesia petersae = Archaeobehnigia edmundsi*) that had been erroneously described by Tshernova (1977) as modern mayfly larvae from the Middle or Upper Jurassic of Transbaikals, can be attributed the order Coxoplectoptera. The discovery of Coxoplectoptera represented one of the more spectacular findings of paleontology in 2011 and was heavily covered by news media around the globe.

Description

Adult

The adult stage of the type species *Mickoleitia longimanus* had a wing length of 28-29 mm and a probable body length of ca. 35-40 mm (the abdomen is not preserved in the single known fossil holotype specimen). A second unnamed species of the genus *Mickoleitia* was only of half this size, and is only known by a single adult specimen from a private fossil collection in Japan. The head of *Mickoleitia* was provided with large compound eyes and functional mouthparts (preserved are 3-segmented labial palps). The thoracic segments are obliquely tilted backwards as in dragonflies, so that the raptorial forelegs are shifted forwards. All legs have a strongly prolonged and free coxal segment. The forelegs are developed as subchelate raptorial devices with a single-segmented tarsus with an unpaired claw. Most likely the abdomen was provided with three caudal filaments (two lateral cerci and the median epiproct) as in modern mayflies and their Permian stem group representatives (Permoplectoptera, e.g. Protereismatidae). Since males of modern mayflies and of Permoplectoptera have gonopods on the 9th abdominal segment that are developed as genital claspers to grip the female for copulation, such a character state and behaviour is also likely for Coxoplectoptera, who have an intermediate position as phylogenetic link between these two groups.

Larva

The more than 20 described larvae of different stages have a body length of 10-32 mm. Their laterally compressed body is unique among all known fossil and Recent aquatic insect larvae, and rather resembles the body of gammarid freshwater shrimps. Many of the fossil larvae are preserved in a characteristic posture with arched back, erect antennae and terminal filaments, and forelegs always in catching position similar to a praying mantis. The head was strongly armored and provided with horn- or shovel-like projections. Of the mouthparts only the crossed, sabre-like mandibles and the spoon-shaped labium are known. All legs have a strongly prolonged and free coxal segment as in the adult. Likewise, the forelegs are developed as slender subchelate raptorial legs with nearly identical segment proportions as in the adult stage, but with a shorter tibia that may have been fused with the single-segmented tarsus, which ended in an unpaired claw. Styliform and ventrally directed abdominal gills are developed on abdominal segments 1-7. These gills are composed of a broader, more strongly sclerotized basal part and a slender and rather membranous distal part. The gills articulate dorsally within the abdominal [Tergum | tergites]] that are distinctly separated from the ventral sternites. The caudal filaments are formed by the two lateral cerci and the slightly longer medial terminal filament. All three appendages are lined with dense rows of long and thin setae.

Ecology and Behaviour

Adult

Because in the adult holotype specimen well-preserved mouthparts (palps) are visible, the adult animals almost certainly have been feeding. On the contrary the adult insect in modern mayflies is just a non-feeding reproductive stage with strongly reduced mouthparts. The raptorial forelegs and oblique thorax indicate that *Mickoleitia* was a predator. The large and broad hinds suggest that they were swift fliers like dragonflies and most probably were actively hunting other flying insects.

Larva

The abundance of fossils, the circumstances of preservation and special anatomical adaptations (7 pairs of abdominal gills, 3 caudal filaments with dense rows of swimming hairs) prove that the larvae have been living in freshwater of streams and rivers, just like those of modern mayflies. They were washed in as allochthonous elements into the brackish Crato lagoon, were the limestones were deposited. The raptorial forelegs, sabre-like mandibles, large eyes and long antennae indicate that the larvae were predators like the adults. On the other hand, the strong, shortened and broadened mid- and hind legs, the strong body armature, and shovel-like projections on the head all suggest that the animals were burrowing. Staniczek, Bechly & Godunko (2011) therefore assumed that the larvae were ambush predators that were hiding, partly burrowed in the [Stream bed|river bed]], and waiting for small prey passing by.

Evolution and Phylogeny

The larvae of Coxoplectoptera provided new clues to the disputed question of the evolutionary origin of insect wings. Before this discovery the paranotal-hypothesis and the leg-exite-hypothesis have been considered as incompatible alternative explanations, which have both been supported by a set of evidences from the fossil record, comparative morphology, developmental biology and genetics. The expression of leg genes in the ontogeny of the insect wing has been universally considered as conclusive evidence in favour of the leg-exite-hypothesis, which proposes that insect wings are derived from mobile leg appendages (exites). However, the larvae of Coxoplectoptera show that the abdominal gills of mayflies and their ancestors, which are generally considered as corresponding structures to insect wings, articulated within the dorsal tergite plates.

This cannot be seen in modern mayfly larvae, because their abdominal tergites and sternites are fused, without any traces of separation left even in embryonic development. If larval gills and wings are corresponding ("serial homologous") structures and thus share the same evolutionary origin, the new results from Coxoplectoptera demonstrate that also wings are of tergal origin, as proposed by the classical paranotal-hypothesis. Staniczek, Bechly & Godunko (2011) therefore suggested a new hypothesis that could reconcile the apparently conflicting evidence from paleontology and developmental genetics: wings first originated as stiff outgrowths of tergal plates (paranota), and only later in evolution became mobile, articulated appendages through secondary recruiting of leg genes.

Within pterygote insects the Coxoplectoptera represent the sister group of modern mayflies (Ephemeroptera). This relationship is indicated by several synapomorphies, such as: adult wing venation with costal brace (absent in other winged insects), larvae with 7 pairs of abdominal gills (compared to still 9 pairs in Permoplectoptera like *Protereisma* larvae), and with single-segmented tarsus with unpaired claw (compared to 3-segmented tarsus with paired claw in Permoplectoptera like *Protereisma* larvae). Together with mayflies and dragonflies they belong to the clade

Palaeoptera, which is characterized by a derived wing articulation with fused sclerites, a vertical resting position of the wings in the groundplan, and a wing venation with intercalary veins between the main longitudinal veins (esp. IR1+ between RP1- and RP2-, and IR2+ between RP2- and RP3/4-).

Because of some very primitive character states, the Coxoplectoptera rather looked like early Paleozoic ancestors of mayflies, e.g. in the wing venation of the adult stage they still had the elongate costal brace that is not fused to the costal margin, and in the larval stage they still had articulated lateral wing pads. The large and broad hind wings are a further plesiomorphy compared to the small hind wing of modern mayflies, and even compared to the slender hind wing of Permian stem group mayflies like *Protereisma*. The monophyly of Coxoplectoptera is demonstrated by several autapomorphic characters in the adult stage, such as the raptorial forelegs and single-segmented tarsi with unpaired claw, as well as in the larval stage by the laterally compressed body, the body armature, the raptorial forelegs and burrowing mid- and hind legs, and the styliform shape of the ventrally directed abdominal gills. Coxoplectoptera are only known from the Jurassic and the Lower Cretaceous. It is not yet known why and when they went extinct.

Systematics

The order Coxoplectoptera only includes a single family Mickoleitiidae with two Mesozoic genera: *Mickoleitia* (Lower Cretaceous, Crato Formation, Brazil):

- *Mickoleitia longimanus* (type species)
- *Mickoleitia spec.* (smaller unnamed species, represented by a single adult specimen in a private fossil collection in Japan)

Mesogenesia (Middle or Upper Jurassic, Transbaikals):

- *Mesogenesia petersae* (= *Archaeobehnigia edmundsi*)

PTERYGOTA

Pterygota is a subclass of insects that includes the winged insects. It also includes insect orders that are secondarily wingless (that is, insect groups whose ancestors once had wings but that have lost them as a result of subsequent evolution). The pterygotan group comprises almost all insects. The hexapod orders not included are the Archaeognatha (jumping bristletails) and the Thysanura (silverfishes and firebrats), two primitively wingless insect orders. Also not included are the three orders that are no longer considered to be insects: Protura, Collembola, and Diplura.

Systematics

Traditionally, this group was divided into the infraclasses Paleoptera and Neoptera. The former are nowadays strongly suspected of being paraphyletic, and better treatments (such as dividing or dissolving the group) are presently being discussed. In addition, it is not clear how exactly the neopterans are related among each other. The Exopterygota might be a similar assemblage of rather ancient hemimetabolous insects among the Neopteras like the Palaeoptera are among insects as a whole. The holometabolous Endopterygota seem to be very close relatives indeed, but nonetheless appear to contain several clades of related orders, the status of which is not agreed upon. The following scheme uses finer divisions than the one above, which is not well-suited to correctly accommodating the fossil groups.

"Infraclass Paleoptera"

(probably paraphyletic)

- Ephemeroptera (mayflies)
- Palaeodictyoptera (Extinct)
- Megasecoptera (Extinct)
- Archodonata (Extinct)

- Diaphanopterodea (Extinct)
- Protodonata or Meganisoptera (Extinct; sometimes included in Odonata)
- Protanisoptera (Extinct; sometimes included in Odonata)
- Triadophlebioptera (Extinct; sometimes included in Odonata)
- Protozygoptera or Archizygoptera (Extinct; sometimes included in Odonata)
- Odonata (dragonflies and damselflies)

Infraclass Neoptera

Superorder Exopterygota

- Caloneurodea (Extinct)
- Titanoptera (Extinct)
- Protorthoptera (Extinct)
- Plecoptera (stoneflies)
- Embioptera (webspinners)
- Zoraptera (angel insects)
- Dermaptera (earwigs)
- Orthoptera (grasshoppers, etc.)
- Proposed superorder Dictyoptera
    - Phasmatodea (stick insects - tentatively placed here)
    - Grylloblattodea (ice-crawlers - tentatively placed here)
    - Mantophasmatodea (gladiators - tentatively placed here)
    - Blattaria (cockroaches)
    - Isoptera (termites)
    - Mantodea (mantids)
- Proposed superorder Paraneoptera
    - Psocoptera (booklice, barklice)
    - Thysanoptera (thrips)

  - Phthiraptera (lice)
  - Hemiptera (true bugs)

Superorder Endopterygota

- Hymenoptera (ants, bees, etc.)
- Coleoptera (beetles)
- Strepsiptera (twisted-winged parasites)
- Raphidioptera (snakeflies)
- Megaloptera (alderflies, etc.)
- Neuroptera (net-veined insects)
- Proposed superorder Mecopteroidea/Antliophora
  - Mecoptera (scorpionflies, etc.)
  - Siphonaptera (fleas)
  - Diptera (true flies)
  - Protodiptera (Extinct)
- Proposed superorder Amphiesmenoptera
  - Trichoptera (caddisflies)
  - Lepidoptera (butterflies, moths)

Neoptera Orders Incertae Sedis

- Glosselytrodea (Extinct)
- Miomoptera (Extinct)

ENTOMOPHAGY

Entomophagy is the consumption of insects as food. Insects are eaten by many animals, but the term is generally used to refer to human consumption of insects; animals that eat insects are known as insectivores, while there are also some species of carnivorous plants that derive nutrients from insects. Human insect-eating is common in cultures in parts of the world, such as North, Central and South America; and Africa, Asia, the Middle East, Australia and New Zealand. Over 1,000 insects are known to be eaten in 80% of the world's nations. However, in some societies insect-eating is uncommon or even taboo. Today insect eating is rare in the developed world, but insects remain a popular food in many developing regions of Latin America, Africa, Asia and Oceania. Some of the more popular insect and arachnids eaten around the world include crickets, cicadas, grasshoppers, ants, a variety of beetle grubs (such as mealworms), the larvae of the darkling beetle or rhinoceros beetle, a variety of species of caterpillar (such as bamboo worms, mopani worms, silkworms and waxworms), scorpions and tarantulas. Entomophagy is sometimes defined broadly to include the practice of eating arthropods that are not insects, such as arachnids (tarantulas mainly) and myriapods (centipedes mainly). There are 1,417 known species of arthropods, including arachnids, that are edible to humans. The term is not used for the consumption of other arthropods, specifically crustaceans like crabs, lobsters and shrimps.

Other Usage

Insects, nematodes and fungi that obtain their nutrition from insects are sometimes termed entomophagous, especially in the context of biological control applications. These may also be more specifically classified into predators, parasites or parasitoids, while viruses, bacteria and fungi that grow on or inside insects may also be termed "entomopathogenic".

History

Before humans had tools to hunt or farm, insects may have represented an important part of their diet. Evidence has been found analyzing coprolites from caves in USA and Mexico. Coprolites in caves in the Ozark Mountains were found to contain ants, beetle larvae, lice, ticks, and mites. This is not unexpected, as there are some deep evolutionary precursors. Firstly, insectivory also features to various degrees amongst primates, such as marmosets and tamarins, and indeed there is some suggestion that the earliest primates were nocturnal, arboreal insectivores. Also, most extant apes are, to a greater or lesser extent, insectivorous. Cave paintings in Altamira, north Spain, dated to about 30,000 to 9,000 BCE, depict the collection of wild bee nests. At the time people must have eaten bee pupae and larvae with the honey. Cocoons of wild silkworm (*Theophilia religiosae*) were found in ruins in the Shanxi province of China, from 2,000 to 2,500 years B.C. The cocoons had large holes, suggesting the pupae were eaten. Many ancient entomophagy practices have been passed down to the present, forming traditional entomophagy.

Current Examples

Entomophagy can be divided into two categories: insects used as a source of nutrients and insects as condiments. Some insects are eaten as larvae or pupae, others as adults. Though not insects, arachnids such as spiders, tarantulas and scorpions are also eaten. A total of 1417 species of insects have been recorded as being eaten by over 3000 ethnic groups. These include 235 species of butterflies and moths, 344 species of beetles, 313 species of ants, bees and wasps as well as 239 species of grasshoppers, crickets and cockroaches, amongst others. Other commonly eaten insects are termites, cicadas and dragonflies. Insects are known to be eaten in 80% of the world's nations. The commercial exploitation of food insects has led to their decline in some places. The consumption of *Atta laevigata* is traditional in some regions of Colombia and northeast Brazil. In southern Africa, a species of moth called *Gonimbrasia belina* is found throughout much of the region; its large caterpillar, the *mopani* or *mopane worm*, is a source of food protein. In Australia, Witchetty grub is considered a source of food amongst the Indigenous population.

Entomophagy has been featured on some reality television shows, such as *Fear Factor*. Barrington Hall, a former student cooperative at U.C. Berkeley held an annual insect banquet for many years until the co-op was closed down in 1990. The New York Entomological Society held a Centennial Banquet on Wednesday, May 20, 1992 at the Explorers Club in New York. The theme for the evening banquet was the use of insects as food. Appetizers and desserts featured insects in their preparations. The keynote speaker was Dr. Gene DeFoliart of the University of Wisconsin who is a leading authority on entomophagy. The Explorers Club itself holds an annual dinner at New York's Waldorf-Astoria Hotel featuring a wide array of unusual dishes including many featuring insects. Theme park operator Six Flags Inc, based in New York, staged a contest as part of a promotion leading up to Halloween in which it also offered customers free entry or line-jumping advantages if they ate a live Madagascar hissing cockroach; the People for the Ethical Treatment of Animals (PETA) opposed the overall promotion. "Insects do not deserve to be eaten alive especially for a gratuitous marketing gimmick," PETA spokeswoman Jackie Vergerio told Reuters.

Advantages

Insects can be a good source of not only protein, but also vitamins, minerals, and fats. For example, crickets are high in calcium, and termites are rich in iron. One hundred grams of giant silkworm moth larvae provide 100 percent of the daily requirements for copper, zinc, iron, thiamin, and riboflavin. Ants can also contain protein depending on the size of the insect. The smaller the species, the greater the chance of it containing minimal or no protein. Grubs of the sago palm weevil (a staple in Papua New Guinea) are laden with unsaturated fat. Many insects contain abundant stores of lysine, an amino acid deficient in the diets of many people who depend heavily on grain.

Minilivestock

The intentional cultivation of arthropods for human food, referred to as minilivestock, is now emerging in animal husbandry as an ecologically sound concept. Minilivestocking suggests that a wide variety of small animals, including arthropods, be reared as nutritious food, the major advantage being that they do not have to be fed on grains thus saving many crop species for human consumption. It is also considered to be much more ecologically friendly than traditional livestocking. Insects generally have a higher food conversion efficiency than more traditional meats, measured as efficiency of conversion of ingested food, or ECI. While many insects can have an energy input to protein output ratio of around 4:1, raised livestock has a ratio closer to 54:1. This is partially due to the fact that feed first needs to be grown for most traditional livestock. Additionally endothermic (warm-blooded) vertebrates need to use a significantly greater amount of energy just to stay warm whereas ectothermic (cold blooded) plants or insects do not. An index which can be used as a measure is the Efficiency of conversion of ingested food to body substance: for example, only 10% of ingested food is converted to body substance by beef cattle, versus 19–31% by silkworms and 44% by German cockroaches. Studies concerning the house cricket (*Acheta domesticus*) provide further evidence for the efficiency of insects as a food source. When reared at 30°C or more and fed a diet of equal quality to the diet used to rear conventional livestock, crickets showed a food conversion twice as efficient as pigs and broiler chicks, four times that of sheep, and six times higher than steers (oxen) when losses in carcass trim and dressing percentage are counted.

Insects reproduce at a faster rate than beef animals. A female cricket can lay from 1,200 to 1,500 eggs in three to four weeks, while for beef the ratio is four breeding animals for each market animal produced. This gives house crickets a true food conversion efficiency almost 20 times higher than beef. For this reason and because of the essential amino acids content of insects, some people, on ecological grounds, propose the development of entomophagy to provide a major source of protein in human nutrition. Protein production for human consumption would be more effective and consume fewer resources than vertebrate protein. This makes insect meat more ecological than vertebrate meat. Insects have attractive qualities for food production besides their high energy efficiency. For example the spatial usage and water requirements are only a fraction of that required to produce the same mass of food with cattle farming. Production of 150g of grasshopper meat requires only very little water, while cattle requires 3290 liters to produce the same amount of beef.

Disadvantages

Toxicity

In general, many insects are herbivorous and less problematic than omnivores. Cooking is advisable, ceteris paribus, since parasites of concern may be present. But pesticide use can make insects unsuitable for human consumption. Herbicides can accumulate in insects through bioaccumulation. For example when locust outbreaks are treated by spraying, people can no longer eat them. This may pose a problem since edible plants have been consumed by the locusts themselves. Cases of lead poisoning after consumption of chapulines were reported by the California Department of Health Services in November 2003. Adverse allergic reactions are also a possible hazard. In the Carnia region of Italy, moths of the Zygaenidae, known for manufacturing hydrogen cyanide precursors in their body, are eaten by children despite their toxicity. The moths concerned are from the brightly coloured day-flying genus Zygaena and its mimic, Syntomis. Cyanogenic glucosides were present in the larvae and imagos (adults) as well as in the ingluvies but in extremely low quantities along with sugar which is found in much higher concentrations. In early summer, Zygaena is very common and easy to catch by hand and the ingluvies serves as a convenient, supplementary source of sugar. Because the ingluvies have a very low cyanogenic content, children can include this resource as a seasonal delicacy at minimum risk.

Cultural Taboo

Within Western culture, entomophagy (barring some food dyes) is seen as taboo. There are some exceptions. Casu marzu, for example, also called casu modde, casu cundhídu, or in Italian formaggio marcio, is a cheese made in Sardinia notable for being riddled with live insect larvae. Casu marzu means "rotten cheese" in Sardinian and is known colloquially as maggot cheese. A scene in the Italian film *Mondo Cane* (1962) features an insect banquet for shock effect. Western avoidance of entomophagy coexists with the consumption of other invertebrates such as crustaceans and mollusks, and is not based on taste or food value. Within Judaism, most insects are not considered kosher, with only a few species of locust being accepted by certain communities. Honey is, however, considered kosher. Some schools of the Islam considers scorpions unclean, but locust consumption is accepted; other prohibit all animals that creep, including insects. The anthropologist Marvin Harris has suggested that the eating of insects is taboo in cultures that have other protein sources that require less work to obtain, such as poultry or cattle, though there are cultures which feature both animal husbandry and entomophagy. Examples can be found in Botswana, South Africa and Zimbabwe where strong cattle-raising traditions co-exist with entomophagy of insects like the mopane worm.

Unintentional Ingestion

In practice, it is not possible to eliminate pest insects from the human food chain. Insects are present in many foods, especially grains. Food laws in many countries do not prohibit insect parts in food, but rather, they limit the quantity. People in rice-eating regions, for example, typically ingest significant numbers of rice weevil (*Sitophilus oryzae*) larvae, and this has been suggested as an important source of vitamins. The Food and Agricultural Organization specifies in the Codex Alimentarius standard for wheat (Codex Standard 152-1985 : Codex Standard for Wheat Flour) that :

> Wheat flour shall be free from abnormal flavours, odours, and living insects. 3.1.3 Wheat flour shall be free from filth (impurities of animal origin, including dead insects) in amounts which may represent a hazard to human health.

According to the U.S. Food and Drug Administration's The Food Defect Action Levels booklet. Contamination on the average of 150 or more insect fragments per 100 grams of wheat flour, or below poses no health hazard. Other example of the maximum permissible levels of insect contamination in food products for humans, contamination below which level, poses no health hazard, are:

Product	*Type of insect contamination*	*Maximum Permissible Level*
Canned sweet corn	Insect larvae (corn ear worms or corn borers)	2 or more 3 mm or longer larvae, cast skins, larval or cast skin fragments, the aggregate length of insects or insect parts exceeds 12 mm in 24 pounds
Canned citrus fruit juices	Insects and insect eggs	5 or more Drosophila and other fly eggs per 250 ml or 1 or more maggots per 250 ml
Wheat flour	Insect filth	Average of 150 or more insect fragments per 100 grams
Frozen broccoli	Insects and mites	Average of 60 or more aphids and/or thrips and/or mites per 100 grams
Hops	Insects	Average of more than 2,500 aphids per 10 grams
Ground thyme	Insect filth	Average of 925 or more insect fragments per 10 grams
Ground nutmeg	Insect filth	Average of 100 or more insect fragments per 10 grams

See source for information on other food products.

ENTOMOPHAGOUS PARASITE

Entomophagous parasites are insects that are parasitic on other insects. Nearly all insects are attacked by one or more insect parasites. This parasitic mode of life is most often confined to the larvae with the adults usually leading free lives. The vast majority of Hymenopterous insect parasites are usually restricted to specific hosts, such as Embidobia (parasitic on Embioptera), Platygaster and Polygnotus (on Cecidomyiidae), Scelio (on locust eggs) and Aphidiuc (on Aphids). Often, species that are hyperparasites are less host-specific.

Parasitoids

Species whose feeding on the host leads to the certain death of the host, typically before the host reached reproduction, are termed parasitoids. In some cases, the larvae feed externally on paralyzed hosts, typically provided by the parent in special brood chambers; this life history is common among Scoliidae, Thynnidae, Mutillidae, and other predatory wasps. Others seek the eggs, larvae or pupae of various insects and deposit their eggs within them. The larvae develop on these stages of the host, killing it in the end. The parasitoid mode of life is common in Hymenoptera, especially within the Ichneumonoidea, Chalcidoidea and Proctotrupoidea and in Diptera (Tachinidae). Strepsiptera differ in that the adult males live free lives with the females remaining as endoparasites, in a larval like state, feeding on the host in many Neuroptera and in the Epipyropidae among Lepidoptera.

Superparasitism

Superparasitism is a form of parasitism in which the host (typically an insect larva such as a caterpillar), is attacked more than once by a single species of parasitoid.

Multiparasitism

Multiparasitism, also called multiple parasitism, on the other hand, occurs when the host has been parasitized by more than one species.

Hyperparasitism

In hyperparasitism, a parasite of an insect is itself attacked by another insect parasite. Species of the chalcid *Perilampus* are often parasitic on the ichneumonoid genera *Microgaster* and *Apanteles*, which are parasites of caterpillars of Lepidoptera. The hyperparasites are usually far less restricted in their selection of host than the parasites. Many hyperparasites are ectoparasites. The effect of attack by the parasite is usually slow death of the host, though very rarely it leads to castration of the host (as in parasitism by Strepsiptera).

Social Parasites

The term social parasitism is given to the relation of the guests in the nests of various social insects. In the case of certain species of ants this takes a special form called slavery. The slave-driving ants regularly capture the pupae from the nest of other species of ants, carry them to their own nest and compel the workers emerging from the pupae to breed their own brood. The ant Anergates indeed rears no workers of its own, but invades the nests of Tetramorium and carries off the brood to its own nests. They are incapable of feeding for themselves but have to be fed by the slaves. Robbing is also a kind of social parasitism called cleptoparasitism. The so-called bee lice invade the nests of bees and devour the store of food in them. Brood parasitism, met within the Chrysididae, Mutillidae, robber bees, parasitic bees and in certain Meloid beetles, refers to the smuggling of the eggs into the brood nest of another insect.

REFERENCES

Bouvresse, S.; Chosidow, O. (Apr 2010). "Scabies in healthcare settings". *Curr Opin Infect Dis* 23 (2): 111–8.

Capinera, John L. (2004). *Encyclopedia of Entomology.* Kluwer Academic Publishers.

Carol Turkington and Jeffrey S. Dover, M.D. (2006). *The Encyclopedia of Skin and Skin Disorders.* New York: Facts on File inc.

Chapman, R.N, (1931). *Animal Ecology*, London and New York.

Christopher O'Toole (2002). *Firefly Encyclopedia of Insects and Spiders.*

Cockerell, T.D.A. (1891). *The Use of Ants to Aphids, and Coccids*, Nature, 44: 608.

Folsom, J. W. (1934). *Entomology with Special Reference to its Ecological Aspects*, Philadelphia: Blackiston and Sons, p. 605.

Gates, Robert H. (2003). *Infectious disease secrets* (2. ed.). Philadelphia: Elsevier, Hanley Belfus, p. 355.

Gordon, David George (1998). *The Eat-A-Bug Cookbook.* Berkely, California: Ten Speed Press.

H. Charles & J. Godfray (2004). "Parasitoids". *Current Biology Magazine* 14 (12): R456.

H.V. Hoell, J. T. Doyen & A.H. Purcell (1998). *Introduction to Insect Biology and Diversity* (2nd ed.). Oxford University Press. pp. 407–409.

Hay R.J. (2009). "Scabies and pyodermas—diagnosis and treatment". *Dermatol Ther* 22 (6): 466–74.

Hoell, H.V., Doyen, J.T. & Purcell, A.H. (1998). *Introduction to Insect Biology and Diversity, 2nd ed.*. Oxford University Press, p. 320.

Lively, Curtis M. and Dybdahl, Mark F. (8 June 2000). "Parasite adaptation to locally common host genotypes." *Nature*. Vol. 405.

Mani. M.S. (1968). *General Entomology*, Oxford and IBS Publishing Co. (New Delhi Bombay Calcutta) 228-229.

Markell, Edward K.; John, David C.; Petri, William H. (2006). *Markell and Voge's medical parasitology* (9th ed.). St. Louis, Mo: Elsevier Saunders.

McElroy, A., & Townsend, P.K. (1989). *Medical Anthropology in Ecological Perspective.*. Boulder, Colorado: Westview Press.

Poinar, G.O. (1986). Entomophagous Nematodes. Pages 95-121, In H. Franz (ed.). Biological Plant and Health Protection, Fortschritte der Zoologie, Bd.32.G. Fischer Verlog, Stuttgart, New York

Price, P.W. (1980). Evolutionary Biology of Parasites. Princeton University Press, Princeton.

Rook, G.A.W. (2007). "The hygiene hypothesis and the increasing prevalence of chronic inflammatory disorders". *Transactions of the Royal Society of Tropical Medicine and Hygiene* 101 (11): 1072–4.

Rózsa L, Reiczigel J, Majoros G. (2000). Quantifying parasites in samples of hosts. *Journal of Parasitology,* 86, 228-232.

Rózsa L. (1997). Patterns in the abundance of avian lice (Phthiraptera: Amblycera, Ischnocera). *Journal of Avian Biology,* 28, 249–254.

Saggers, S., & Gray, D. (1991). *Aboriginal Health & Society: The Traditional and Contemporary Aboriginal Struggle for Better Health.*. Sydney: Allen & Unwin.

Shaffie, S.A., A. Alam and M.M. Agrawal, (1955). *Taxonomic Survey of Eneyrtid Parasites in India*, Aligarh Mus. Univ. Pub. Zool) 9.10; p. 238.

Weiss, M.L., & Mann, A.E. (1985). *Human Biology and Behaviour: An Anthropological Perspective.*. Boston: Little Brown & Co.

5

Insect Biodiversity, Measurement Methods and Insect Migration

INSECT BIODIVERSITY

Insect biodiversity accounts for a large proportion of all biodiversity on the planet, with over 1,000,000 insect species described.

Agriculture

In agricultural ecosystems, biodiversity is instrumentally important not only for the production of food, but for other ecological services as well, including the recycling of nutrients, regulation of microclimate and local hydrological processes, suppression of undesirable organisms and detoxification of noxious chemicals. In the United States alone, pollination by bees accounts for over US$9 billion of economic revenue . According to some estimates, over S& of the human diet can be traced directly or indirectly to bee pollination . Losses of key pollinators have been reported in at least one region or country on every continent except Antarctica, which has no pollinators. The Millennium Ecosystem Assessment concluded that with the global decline in the amount of pollinators, there is not a complete loss of fruit or seeds, but a significant decrease in quantity and viability in fruits, and a lower number of seeds.

Food

Over 1000 species of insects are, or have been used as

food somewhere in the world. Entomophagy is common in central and southern Africa, Asia, Australia, and Latin America. Termites, crickets, grasshoppers, locusts, beetles, ants, bee brood, and moth larvae are examples of insects that are used as food sources. Insects are high in protein, energy, and a number of minerals and vitamins and can form up to 5-10% of the annual animal protein consumption of certain indigenous groups . Many insects are said to have a nutty flavour, along with a high nutritional content.

Human Culture

Early human civilizations held insects as an important element within their cultures. Most famously, scarab beetles were central religious artifacts within Egyptian culture. Insect symbolism, such as the dung beetle portrayed as a potter, is seen in the East as well. The Chinese viewed cicadas as a symbol of birth or immortality, the San of the Kalahari believe that the praying mantis represents creation and patience, and the Greeks also created beautiful representations of scarab beetles using colourful stones.

The Scope

Over 1 million species of insects have been described, but current estimates of total insect diversity vary from 5-80 million species of insect. Beetles (Coleoptera) make up 40% of described insect species, but some entomologists suggest that flies (Diptera) and Hymenoptera (wasps, bees and ants) could be as diverse or more so. Five orders of insects stand out in their levels of species richness: Hymenoptera, Diptera, Coleoptera, Lepidoptera, and the Hemiptera.

Conservation

There are two approaches to the conservation of insects. Either humans set aside large portions of land using "wilderness preservation" as the motive, or confronting the particular processes that affect the charismatic vertebrates in order to achieve indirect conservation of insects. With biodiversity loss being a global problem, conserving habitat simply for species of insects is of low priority in the current environmental culture. Single-species conservation is said to preserve many other species indirectly, this preservation by default is referred to as the umbrella effect. "Charismatic species", such as butterflies or large, colourful beetles, called flagship species, can expand public awareness and financial contributions for conservation efforts. Migratory species, such as the well-known monarch butterfly (*Danaus plexippus*), are in need of special conservation methods. One species may require several habitat locations for different periods in their migratory patterns. Insect conservation has been labeled in the past as a concern only for the affluent. The developing country of Papua New Guinea has a "happily ever after" ending in their attempts to preserve the world's largest butterfly, Queen Alexandra's Birdwing (*Ornithoptera alexandrae*). This species is restricted to a very small range of habitat due to specificity in their diet. In the international market of insect collecting, the butterfly can retrieve up to US$2000. In 1978, the government of Papua New Guinea set up the Insect Farming and Trading Agency (IFTA) to regulate the exploitation and conservation of Queen Alexandra's Birdwing and other valuable butterflies.

INSECT COLLECTING

Insect collecting is the collection of insects for hobby, scientific study or profit. Historically insect collecting has been widespread and a very popular educational hobby. Insect collecting has left traces in European cultural history, literature and songs (e.g., Georges Brassens: "La chasse aux papillons" ("The hunt for butterflies"). The practice is still widespread in Japan, especially amongst the youth demographic. Since most types of insects have hard exoskeletons that retain their appearance after the insects dies, it is easy and practical to form a collection. With many thousands of colorful and attractive species known, a collection may become quite large. The collecting of butterflies is perhaps most popular, with beetles in second place.

Nowadays some collectors practice their hobby also by rearing caterpillars, or ova, to adulthood. When done well, this procreates more adults than in nature. Some of the adults are released back in the wild, thus minding the numbers of individuals in natural populations. This method has also been used in some conservation programmes.

Process

Locating Insects

Insects are common in almost every part of world, and may be easily found by looking closely at plant leaves or flowers, under rocks and boards, in water, and so forth.

Collecting Techniques

Aspirators or "pooters" suck up insects too small or delicate to handle with fingers. At night, many types can be caught flying around lights. Funnels, malaise traps, and other types of insect traps, some of which are baited with small bits of sweet foods (such as honey), are also quite effective. Several different types of nets are commonly used to collect insects. Butterfly nets are used to collect flying insects. The bag of a butterfly net is generally constructed from a lightweight mesh to minimize damage to delicate butterfly wings. A sweep net is used to collect insects from grass and brush. It is similar to a butterfly net, except that the bag is generally constructed from more rugged material such as sailcloth. Other types of nets used for collecting insects include beating nets and aquatic nets. Once collected, a killing jar is often used to dispatch insects before they damage themselves trying to escape. However, Killing jars are generally only used on hard-bodied insects.

Soft-bodied insects, such as those in the larval stage, are generally drowned in a vial containing an ethanol and water solution or formaldehyde, which both kills and then preserves them. Butterflies and moths are not traditionally killed in a killing jar as their wings, which are delicate, can become damaged trying to escape. Instead, the usual method of killing a lepidopteran is by crushing the thorax, and death occurs by suffocation.

Display

The usual method of display is in a glass-covered box, with the insects mounted on specially made non corrosive insect pins stuck into foam board (typically Plastazote) or paper covered cork at the bottom of the box. Common pins are not used. Very small insects are either pinned on "minuten" (very tiny pins) into a block of plastazote borne by a standard insect pin or glued to a small piece of card on the pin. There are specific procedures for proper mounting that are used to show off the insects unique characteristics. Sometimes variations can be used especially if more than one of a species is available. For example, one or both of the wings of a beetle or grasshopper can be pulled open and fanned out to show the wing structure that would otherwise be hidden. At least the date and place of capture should be written or computer printed on to a piece of paper or card transfixed by the pin. This is called a data label. Museums frequently have quite large insect collections, used for research. These are not usually on display, since most of them will all look the same to the untrained eye. Rare insects, and/or those from distant parts of the world may also be acquired from dealers or by trading. Some noted insect collections have been sold at auction.

INSECT TRAP

Insect traps are used to monitor or directly reduce insect populations. They typically use food, visual lures, chemical attractants and pheromones as bait and are installed so that they do not injure other animals or humans or result in residues in foods or feeds. Visual lures use light, bright colors and shapes to attract pests. Chemical attractants or pheromones may attract only a specific sex.

Insect traps are sometimes used in pest management programmes instead of pesticides but are more often used to look at seasonal and distributional patterns of pest occurrence. This information may then be used in other pest management approaches. The trap mechanism or bait can vary widely. Light traps with ultraviolet attract certain insects. Designs differ according to the behaviour of the insects being studied. Grasshoppers and some beetles are attracted to lights at a long range but are repelled by it at short range. Farrow's light trap has a large base so that it captures insects that may otherwise fly away from regular light traps. Flies and wasps are attracted by proteins. Mosquitoes and many other insects are attracted by bright colors, carbon dioxide, lactic acid, floral or fruity fragrances, warmth, moisture and pheromones. Synthetic attractants like Methyl eugenol are very effective with Tephritid flies. Yellow pan traps are used to monitor aphids and some other sap sucking insects. Pitfall traps are used for ground foraging and flightless insects such as beetles of the family Carabidae.

Insect Traps in Practice

At present, several traps are used both in the commercial sector (e.g. commercial greenhouses) and by individual gardeners. The commercial sector primarily focuses on highly effective traps as pheromone traps, while gardeners tend to stay with low-tech, cheap traps based on foods or home-grown production.

Examples of Food Traps

Following traps are used extensively by gardeners:

- eggshells: strewn on the ground; protects plants against snails
- sunflower-flowerbottoms: against mice and rats
- carrot disks: against click beetles and crane flies
- sawdust: against onion fly, carrot fly and cabbage flies

- flour grains: made from 3 parts flour, 2 parts chalc, and 1 part sugar, knid to a grain; against mice
- beer and regular pot traps: small pots, filled with 2 cm beer or without anything at all, dug into the soil; against snails and other insects

INSECT MIGRATION

Insect migration is the seasonal movement of insects, particularly those by species of dragonflies, beetles, butterflies and moths. The distance can vary from species to species, but in most cases these movements involve large numbers of individuals. In some cases the individuals that migrate in one direction may not return and the next generation may instead migrate in the opposite direction. This is a significant difference from bird migration. The most famous insect migration is that of the Monarch butterfly which migrates from southern Canada to wintering sites in central Mexico where they spend the winter. In the late winter/early spring, the adult monarchs leave the Transvolcanic mountain range in Mexico for points North. Mating occurs and the females begin seeking out milkweed to lay their eggs, usually first in northern Mexico and southern Texas. The caterpillars hatch and develop into adults that move north, where more offspring can go as far as Central Canada until next migratory cycle.

Definition

All insects move to some extent. The range of movement can vary from within a few centimeters for some sucking insects and wingless aphids to thousands of kilometres in the case of other insects such as locusts, butterflies and dragonflies. The definition of migration is therefore particularly difficult in the context of insects. A behaviour oriented definition proposed is

> Migratory behaviour is persistent and straightened-out movement effected by the animal's own locomotory exertions or by its active embarkation on a vehicle. It

> depends upon some temporary inhibition of station-keeping responses but promotes their eventual disinhibition and recurrence.
>
> —*Kennedy, 1985*

This definition disqualifies movements made in the search of resources and which are terminated upon finding of the resource. Migration on the other hand involves longer distance movement and these movements are not affected by the availability of the resource items.

General Patterns

Migrating butterflies fly within a boundary layer, with a specific upper limit above the ground. The air speeds in this region are typically lower than the flight speed of the insect. These 'boundary-layer' migrants include the larger day-flying insects, and their low-altitude flight is obviously easier to observe than that of most high-altitude windborne migrants. Many migratory species tend to have polymorphic forms, a migratory one and a resident phase. The migratory phases are marked by their well developed and long wings. Such polymorphism is well known in aphids and grasshoppers. In the migratory locusts, there are distinct long and short-winged forms. Migration being energetically costly has been studied in the context of life-history strategies. It has been suggested that adaptations for migration would be more valuable for insects that live in habitats where resource availability changes seasonally. Others have suggested that species living in isolated islands of suitable habitats are more likely to evolve migratory strategies. The role of migration in gene flow has also been studied in many species.

Orientation

Migration is usually marked by well defined destinations which need navigation and orientation. A flying insect needs to make corrections for crosswinds. It has been demonstrated that many migrating insects sense windspeed and direction and make suitable corrections. Day-flying insects primarily make use of the sun for orientation, however this requires that they compensate for the movement of the sun. Endogenous time-compensation mechanisms have been proposed and tested by releasing migrating butterflies that have been captured and kept in darkness to shift their internal clocks and observing changes in the directions chosen by them. Some species appear to make corrections while it has not been demonstrated in others.

Most insects are capable of sensing polarized light and they are able to use the polarization of the sky when the sun is occluded by clouds. The orientation mechanisms of nocturnal moths and other insects that migrate have not been well studied, however magnetic cues have been suggested in short distance fliers. Recent studies suggest that migratory butterflies may be sensitive to the Earth's magnetic field on the basis of the presence of magnetite particles. In an experiment on the monarch butterfly, it was shown that a magnet changed the direction of initial flight of migrating monarch butterflies. However this result was not a strong demonstration since the directions of the experimental butterflies and the controls did not differ significantly in the direction of flight.

Lepidoptera

Migration in the butterflies and moths are particularly well known. The Bogong moth is a native insect of Australia that is known to migrate to cooler climates. The Madagascan sunset moth (*Chrysiridia rhipheus*) has migrations of up to thousands of individuals occur between the eastern and western ranges of their host plant, because they become depleted or unsuitable for consumption. In southern India, mass migrations of many species are noted prior to the monsoons. As many as 250 species of butterflies in India are migratory. These include members of the Pieridae and Nymphalidae.

REFERENCES

Baker, R.R. (1987). Integrated use of moon and magnetic compasses by the heart-and-dart moth, *Agrotis exclamationis*. Animal Behaviour 35, 94–101.

Catala, R. (1940). "Variations expérimentales de *Chrysiridia madagascariensis* Less. (Lep. Uraniidae)". *Archives du Muséum National D'Histoire Naturelle* 17: Ph.D. Thesis.

Chapman, R.F. (1998). *The Insects : Structure and Function*. 4th Edition. Cambridge University Press.

Dickinson, MH; Lehmann, FO; Sane, S.P. (1999). "Wing rotation and the aerodynamic basis of insect flight". *Science* 284 (5422): 1954–60.

Ellington, C.P. (1984). "The Aerodynamics of Hovering Insect Flight. I. The Quasi-Steady Analysis". *Philosophical Transactions of the Royal Society B: Biological Sciences* 305 (1122): 1–15.

Ellington, Charles P.; Van Den Berg, Coen; Willmott, Alexander P.; Thomas, Adrian L.R. (1996). "Leading-edge vortices in insect flight". *Nature* 384 (6610): 626–630.

Ellington, CP (1999). "The novel aerodynamics of insect flight: Applications to micro-air vehicles" (PDF). *The Journal of experimental biology* 202 (Pt 23): 3439–48.

Lewin, G. C.; Haj-Hariri, H. (2003). "Modelling thrust generation of a two-dimensional heaving airfoil in a viscous flow". *Journal of Fluid Mechanics* 492: 339–362. Bibcode 2003 JFM...492..339L.

Lighthill, M.J. (2006). "On the Weis-Fogh mechanism of lift generation". *Journal of Fluid Mechanics* 60: 1. Bibcode 1973JFM....60....1L.

Perez, S.M., Taylor, O.R. and Jander, R. (1999). The effect of a strong magnetic field on monarch butterfly (*Danaus plexippus*) migratory behaviour. Naturwissenschaften 86, 140–143.

S.E. McGregor (1976). *Insect pollination of cultivated crop plants*. USDA Agriculture Handbook 496, Washington, D.C.

W.S. Robinson, R. Nowogrodski & R.A. Morse (1989). "Pollination parameters". *Gleanings in Bee Culture* 117: 148–152.

6

Other Major Areas of Study of Entomology: A Select Overview

ENTOMOLOGICAL EVIDENCE COLLECTION

Entomological evidence collection is the process of collecting evidence based on insect clues to be used in criminal investigations. If evidence is not carefully preserved at a crime scene after a death, it may be difficult or impossible for an entomologist to make an accurate identification of specimens, if for example, all morphological characteristics are not preserved.

Collecting Materials

One of the most valuable things a forensic entomologist has is his collection kit. There are commercially available insect evidence collection kits, but you can customize your own to meet geographical and budget requirements. Below is a list of necessary tools.

- Aerial or sweep insect nets: Nets can be anywhere from 15-36"
- Sticky Traps:
- Feather weight or light touch forceps
- Collection Vials: Screw cap type with neoprene cap inserts 4-dram size, killing jars which are jars of ½" of plaster to absorb the ethyl acetate
- Plastic "yogurt" or "bait" containers: Used in collecting and shipping larvae

- Aluminum foil: Used in constructing a pouch that will hold the live larvae and their food source during shipment
- Vermiculite: Used in filling the bottom of the larval containers
- Plastic specimen containers 4-8oz size
- Non-adhesive and heavy bond paper labels used in labeling and placement inside and outside of the collection containers
- #2 pencil: Used in making labels
- Small hand trowel or garden spade: Used in soil sampling and digging for migrating larvae or pupae in outdoor death scenes
- Thermometers: Preferably digital, but dial type, mercury, or alcohol types are fine
- 35 mm SLR camera, lens, and flash: Used for on the scene macrophotography of insects. Also, a standard 35 mm lens should be used for general scene photographs.
- Ruler or any other measuring device: For showing scale in photographs
- Disposable gloves
- Preservation and collection chemicals: In particular ethyl alcohol, ethyl acetate, and KAA
- Paper towels: For use in kill jars and for cleaning utensils
- Thermos of boiling water: Used to boil the larvae
- Sifting screens: Used for processing soil samples for insects and insect artifacts
- Entomological Death Scene Form
- Shipping containers: Styrofoam containers with lids are best as it offers good insulation from temperature extremes, but corrugated cardboard boxes are widely used
  - This list was adapted from Dr. J.H. Boyd's website for entomological collection equipment and materials.

Collecting Specimens

In order for entomological evidence to be of value for a death scene investigation it is important that every detail be noted and preserved. Great care must be exercised when collecting insects to ensure that morphological characteristics are kept intact for positive species identification. It is also important to make detailed observations of the bodies surroundings so that inferences to the events preceding death can be made by linking the stages of insects found to their access to the body. This is accomplished by using the proper tools and paper work to collect and record all available evidence. In most cases, it is important to collect samples for the study of insect succession three times a day for about a month. Environmental and body temperatures also have to be taken into account. when determining the post mortem interval.

Collecting Insects on the Body

Great care should be used when removing insects from the body prior to it being removed from the scene. During this collection the body, clothing, and immediate surroundings must not be disturbed. Insects may only be removed from the surface of the body, and great care must be taken with forceps and other tool to prevent postmortem artifacts. Any damage to the body or integrity of the scene can create misleading evidence that could be detrimental to the investigation. Also, the location on the body were each sample of specimens was collected should be clearly noted on the specimens label.

Collecting Eggs

Eggs are extremely small and delicate, so great care should be taken in their collection. Eggs are generally laid in clumps or masses, and are found in any orifice whether it be a natural opening or a wound. The safest way to remove eggs is by using a small paint brush dipped in water, so that the eggs stick to the bristles. Forceps may be used but can often damage the eggs. Half of the eggs should be placed into a preservative, while the other half should be kept alive to be reared out. Since it is often difficult to identify insect eggs rearing the eggs to adulthood provides positive identification. In order to prevent the eggs from dehydrating it is important to place a damp tissue or other media in the collection vial. It also important to make sure the eggs have air, placing a paper towel over the top using a rubber band provides plenty of ventilation. Newly emerged maggots are very small so the vial must remain upright to prevent escape.

Collecting Maggots and Larva

It is important to collect all the various sizes of maggots found at each maggot mass. Larger maggots usually represent the first flies to arrive on the body, but different sized maggots can also represent different species. As mentioned earlier it is important to note the maggot mass location and temperature, because this will influence the degree days it takes for the fly to develop. It is important to preserve the specimen to see their exact state of development for evidences later on. This can be done by boiling the maggots, and then placing them in a 70% alcohol or 50% isopropyl alcohol. A sample of living specimens approximately 100 of each size or distinguishable species, should be kept in a maggot rearing container. It is important to keep the species and sizes separate to prevent cannibalism. Since 1st and 2nd instars can be delicate it may be necessary to remove them with a brush in the same manner discussed with the eggs. The more robust 3rd instar maggots may be removed with forceps (Anderson). Another technique that may be used for removing maggots is to scoop them up with plastic spoon.

Pupa

Pupa can occasionally be found on the clothing of the body, and may be removed using forceps or fingers. Since pupa will not grow, it is not necessary to preserve them. If a specimen is found with a pale color, it has just entered pupation. Such a specimen should be kept separate because its exact age can be determined accurately, to within a few hours. All insects found on the body should be placed into kill jars and properly labeled. For the most part adult flies are only important for indicating what species of fly may develop on the corpse. However, if an adult with crumpled wings is found it should be labeled as such, because this indicates that it may have just emerged from its pupil casing. This indicates that at least one generation of flies has developed off the corpse. In addition, sticky traps can also be used by placing them around the body in order to collect the insects moving around the body (mainly flies). You will usually place the entire sticky trap underneath the microscope in order to identify the species of the insects on the trap, because it is very difficult to separate them off the trap. Many other insect besides flies are attracted to the scene because of the odor, while others are just in the wrong place at the wrong time. You cannot collect all the insects on the scene and, therefore, you have to know which insects to collect. One other insect that is forensically important is the beetle. This insect is known to provide clues to the degree of exposure the body has experienced. Wearing gloves are good for caputuring adult beetle and larvae. You are able to transfer both the adult and the larvae beetle into a proper collection container using your hands without the worry of crushing them because of their sclerotized exoskeleton.

Techniques for Collecting Specimens Before Body Removal

Pup Tent Fly Traps

This technique was developed in Knoxville, Tennessee in 1989, and in the pilot study was found to be more effective than using an aerial net. The traps are constructed using manufactured sticky traps measuring approximately 9 inches (230 mm) by 10 inches (250 mm) in the shape of an inverted "V", or pup tent, with the sticky material on the outside. These traps generally have four clothespins clipped to the bottom corners to provide ground clearance. The traps are placed at varying distances from the body, between 1 foot (0.30 m) and 12 feet (3.7 m), with one side facing the body. The sides of the traps should be set at a 60 degree angle. When collection is done the sticky trap should be labeled with both the time it was put out and the time it was picked up. In the lab insect can be individually removed and rinsed with paint thinner or xylene before preservation.

Sweeping

Generally insects associated with carrion are strong and fast fliers, but with practice can be caught using an aerial net. The sweeping technique involves making quick back and forth sweeps across the body in succession. Each time direction is changed the mouth of the net must be flipped 180 degrees to maintain one entrance. Speed is important not only for capturing insects, but for keeping those already caught inside. Generally 6-10 sweeps are sufficient, and on the last sweep the opening of the net is lifted so that is horizontal at about chest height, and then flipped 180 degrees to close the net. The tail end of the net can then place into a kill jar with the lid closed. After about 2 to 5 minutes the insects can be preserved. This procedure should be repeated approximately 3 times to ensure a good sample of flying insects in the area. Note that it may be necessary to use a sturdier net it wooded or brushy areas, so that the net won't tear.

Labeling

As soon as a specimen is collected it should be properly labeled. Labels should include all of the following sample number, exact time, date, case number, and county/city. It is important that a label be place inside the sample vial and on the out side as well. Note that the label should never be placed on the lid, since lids can be easily switched. It is also imperative to never use pen, a graphite pencil must be used at all times. Many of the preservatives used can make ink run, and that not only destroys the label but can damage the specimen. When labeling specimens collected from the body it is important to note the exact location from which they were removed.

Procedure after the Removal of the Body

After the removal of the body from the death scene, there are procedures to be followed in order to obtain an accurate time of death. One of the methods of documentation is the death scene case study form. This form is used to record the condition of the body as well as the surroundings of where the body was found. Before evidence is collected, photographs will be taken in order to document the scene.

Under Body Procedures

Immediately after the body is removed from the scene, the soil which was under the body is observed. Examining the soil will produce evidence to help in the investigation. Under the body there can be beetles and maggots burying themselves in the soil. This is why it is important to take soil samples directly under the body so as to recover these insects that might be hidden from plain sight. Insects of all stages of development are collected. These stages can be in the form of eggs, larva (of which the largest instars are the most important for identification) pupa and adults. Half of these insects will be preserved and the other half will be reared to adults for analysis in the lab. Information on other insects found in the area that are known to be predaceous is documented. The predacous insects could be responsible for insects that are usually present on a body. Litter under the body and the litter in close proximity of where the body was discovered will be collected. These litter samples could hold insects too small to collect. In order to separate the insects from the litter a Berlese/ Tullgren funnel is used.

Temperature, Weather and Climate

Temperature, weather and climate are important readings to record. These readings will aid in determining degree days. An insect's development is based on these factors. There are a few different temperature readings that are taken of the area. The first is the air temp at chest height. Secondly, the temperature between the body and the ground is recorded. The next temperature to take is the ground surface immediately under the body, 1-2 M away from the body which includes three different sources. These three sources are under the grass or ground litter, 4 cm soil depth and 20 cm soil depth (http://brazoria_county.com). The maggot mass temperature will be recorded. The weather and climate data is collected from two different sources. The first source will be from the scene of death itself. Maximum and minimum temperatures for three to five days will be recorded. Accuracy in collecting the weather and climate data is important. Conditions such as direct sunlight or dewfall can affect the accuracy of the data taken. The instrument used to record relative humidity is the psychrometer. This data will then be compared to the data collected from the second source which would be the National Weather Service (NWS Station) or an approved agency. The data required from the NWS Station is for the time period of two weeks before the estimated time of death to three to five days after the estimated time of death. The use of the recordings from the forensic entomologist of the death scene temperatures and the NWS station combined will increase the accuracy of the post mortem interval.

Analysis of Area

Use of the death scene case study form assists in documenting the immediate area of where the body was discovered. While analyzing the area the method of concentric circles is utilized in order to ensure evidence is not overlooked. While walking in these circles, it is important to note major infestations of the insects in an area of twenty feet of where the body was located. This area can contain the different insects that were once on the body and have left due to the movement of the body. Track marks left in the ground left by large masses of maggots leaving the body in order to pupate can be identified by an entomologist. With the presence of an entomologist at the scene, evidence such as this can be correctly identified instead of investigators inaccurately determining the body was drug across the ground. The death scene case study form includes recording whether the area is rural, urban, or aquatic. Recording where the body was found is also important to note as decomposition happens at different rates depending on where the body was located. A body in the air will decompose faster than a body in the water and the body in the water will decompose faster than one found buried in the ground. This information is compiled in order to determine the postmortem interval.

Current Research

Currently, research is being done to determine when a body was placed outside, either day or night. For example, by comparing developmental stages of Lucilia and Calliphora, the time of day should be able to be distinguished. Calliphora is thought to not oviposit at night while Lucilia does, so when evaluating the late stages of Lucilia and the beginning stages of Calliphora, time can be theorized. Another area of research, entomotoxicology Entomotoxicology, is understanding what drugs, either street or prescribed, does to insects when consuming a corpse that has ingested the drug. Also, whether the drug deters some insects or attracts them to the body. An advancement that has recently been made is due to the development of techniques such as the polymerase chain reaction, PCR, to identify remains. Researchers have been able to isolate the gut content of insects found on bodies and use techniques similar to PCR to recover mitochondrial DNA. This mitochondrial DNA has been used to identify the species of the insect as well as the human remains. Advancements in technology such as this will help solve cases, especially during the late decomposition stage. As shown in the case studies provided, the preservation and correct analysis of the various insects found in crime scenes is of the utmost importance. There are various tools and methodologies used for interpreting evidence to make accurate assumptions. Forensic entomology has evolved over the years, and with the discovery techniques becoming more sophisticated which has led to an increase in the uncovering of the truths in multiple murders.

ETHNO-ENTOMOLOGY

Ethnoentomology is the study of the relationship between insects and people. The name is derived from "ethno" - study of people and "entomology" - study of insects. The focus of ethnoentomology is on how insects have been or are being used in human societies around the world. This includes insects used for food, rituals and medicine.

Insects as Food

Entomophagy refers to the eating of insects. Many insects are considered a culinary delicacy in many parts of the world, but the practice is uncommon and even taboo in other societies. Insects used in food include caterpillars, silkworms, Maguey worms, Witchetty grubs & other beetle and moth larvae; crickets, grasshoppers & locusts; and arachnids, such as spiders & scorpions. They can be eaten on their own or mixed with other ingredients, such is the case with casu marzu.

Insects in Medicine

One of the most commonly used insects in medical purposes is the blow fly larvae. The first recorded use of it was during World War II. Military surgeons noticed that wounds which were left untreated for several days and which became infested with maggots, healed better than wounds not infested with the blow fly larvae. It was later discovered that the larvae secreted a chemical called allantoin, which had a curative effect. Allantoin is now being used to treat the infectious bone disease, Osteomyelitis. Maggot Debridement therapy is the intentional introduction of live, disinfected fly larvae into non-healing or dead skin and soft tissue wounds of a human or other animal for the purpose of selectively cleaning out only the necrotic tissue within a wound in order to promote wound healing. It is also used to prevent infection and to speed the healing process. Apitherapy is the medical use of honeybee products. One of the major peptides in bee venom, called Melittin, is used to treat inflammation in sufferers of Rheumatoid arthritis and Multiple sclerosis. Melittin blocks the expression of inflammatory genes, thus reducing swelling and pain. It is administered by direct insect sting, or intramuscular injections Cantharidin, the blister-causing oil found in several families of beetles, was accepted by the FDA in 2004 as treatment for warts and other skin problems. It also has historical use by the Greeks and Romans and is used as an aphrodisiac in some societies. Another extract from Blister beetles has also been used to treat problems of the urogenital system. In the Brazilian Amazon, members of the Tupí–Guaraní language family have been observed using *Pachycondyla commutata* ants during female rite-of-passage ceremonies, and prescribing the sting of *Pseudomyrmex* spp. for fevers and headaches.

ENTOMOLOGICAL WARFARE

Entomological warfare (EW) is a type of biological warfare that uses insects to attack the enemy. The concept has existed for centuries and research and development have continued into the modern era. EW has been used in battle by Japan and several other nations have developed and been accused of using an entomological warfare programme.

Description

Entomological warfare (EW) is a specific type of biological warfare (BW) that uses insects in a direct attack or as vectors to deliver a biological agent, such as plague or cholera. Essentially, EW exists in three varieties. One type of EW involves infecting insects with a pathogen and then dispersing the insects over target areas. The insects then act as a vector, infecting any person or animal they might bite. Another type of EW is a direct insect attack against crops; the insect may not be infected with any pathogen but instead represents a threat to agriculture. The final method of entomological warfare is to use uninfected insects, such as bees, to directly attack the enemy.

Early History

Entomological warfare is not a new concept; historians and writers have studied EW in connection to multiple historic events. A 14th century plague epidemic in Asia Minor that eventually became known as the Black Death is one such event that has drawn attention from historians as a possible early incident of entomological warfare. That plague's spread over Europe may have been the result of a biological attack on the Crimean city of Kaffa. According to Jeffrey Lockwood, author of *Six-Legged Soldiers* (a book about EW), the earliest incident of entomological warfare was probably the use of bees by early humans. The bees or their nests were thrown into caves to force the enemy out and into the open. Lockwood theorizes that the Ark of Covenant may have been deadly when opened because it contained deadly fleas. During the American Civil War the Confederacy accused the Union of purposely introducing the harlequin bug in the South. These accusations were never proven, and modern research has shown it more likely that the insect arrived by other means. The world did not experience large-scale entomological warfare until World War II; Japanese attacks in China were the only verified instance of BW or EW during the war. During, and following, the war other nations began their own EW programmes.

World War II

Canada

Among the Allied Powers, Canada led the pioneering effort in vector-borne warfare. After Japan became intent on developing the plague flea as a weapon, Canada and the United States followed suit. Cooperating closely with the United States, Dr. G.B. Reed, chief of Kingston's Queen's University's Defense Research Laboratory, focused his research efforts on mosquito vectors, biting flies, and plague infected fleas during World War II. Much of this research was shared with or conducted in concert with the United States. Canada's entire bio-weapons programme was ahead of the British and the Americans during the war. The Canadians tended to work in areas their allies ignored; entomological warfare was one of these areas. As the U.S. and British programmes evolved, the Canadians worked closely with both nations. The Canadian BW work would continue well after the war, including entomological research.

France

France is known to have pursued entomological warfare programmes during World War II. Like Germany, the nation suggested that the Colorado potato beetle, aimed at the enemy's food sources, would be an asset during the war. As early as 1939 biological warfare experts in France suggested that the beetle be used against German crops.

Germany

Germany is known to have pursued entomological warfare programmes during World War II. The nation pursued the mass-production, and dispersion, of the Colorado potato beetle (*Lepinotarsa decemlineata*), aimed at the enemy's food sources. The beetle was first found in Germany in 1914, as an invasive species from North America. There are no records that indicate the beetle was ever employed as a weapon by Germany, or any other nation during the war. Regardless, the Germans had developed plans to drop the beetles on English crops. Germany carried out testing of its Colorado potato beetle weaponization programme south of Frankfurt, where they released 54,000 of the beetles. In 1944, an infestation of Colorado potato beetles was reported in Germany. The source of the infestation is unknown, speculation has offered three alternative theories as to the origin of the infestation. One option is Allied action, an entomological attack, another is that it was the result of the German testing, and still another more likely explanation is that it was merely a natural occurrence.

Japan

Japan used entomological warfare on a large-scale during World War II in China. Unit 731, Japan's infamous biological warfare unit, used plague-infected fleas and flies covered with cholera to infect the population in China. The Japanese military dispersed the insects by spraying them from low-flying airplanes and dropping bombs filled with a mixture of insects and disease. Localized and deadly epidemics resulted and nearly 500,000 Chinese died of disease. An "international symposium" of historians declared in 2002 that Japanese entomological warfare in China was responsible for the deaths of 440,000.

United Kingdom

A British scientist, J.B.S. Haldane, suggested that Britain and Germany were both vulnerable to entomological attack via the Colorado potato beetle. In 1942 the United States shipped 15,000 Colorado potato beetles to Britain for study as a weapon.

Cold War

Soviet Union

The Soviet Union researched, developed and tested an entomological warfare programme as a major part of an anti-crop and anti-animal BW programme. The Soviets developed techniques for using insects to transmit animal pathogens, such as foot and mouth disease — which they used ticks to transmit. The nation also used avian ticks to transmit *Chlamydophila psittaci* to chickens. In addition, the Soviet Union claimed to have developed an automated mass insect breeding facility, capable of outputting millions of parasitic insects per day.

United States

The United States seriously researched the potential of entomological warfare during the Cold War. The United States military developed plans for an entomological warfare facility, designed to produce 100 million yellow fever-infected mosquitoes per month. A U.S. Army report titled "Entomological Warfare Target Analysis" listed vulnerable sites within the Soviet Union that the U.S. could attack using entomological vectors. The military also tested the mosquito biting capacity by dropping uninfected mosquitoes over U.S. cities. North Korean and Chinese officials leveled accusations that during the Korean War the United States engaged in biological warfare, including EW, in North Korea. The claim is dated to the period of the war, and has been thoroughly denied by the U.S. In 1998, Stephen Endicott and Edward Hagermann claimed that the accusations were true in their book, *The United States and Biological Warfare: Secrets from the Early Cold War and Korea* The book received mixed reviews, some called it "bad history" and "appalling", while other praised the case the authors made. Other historians have revived the claim in recent decades as well. The same year Endicotts' book was published Kathryn Weathersby and Milton Leitenberg of the Cold War International History Project at the Woodrow Wilson Center in Washington released a cache of Soviet and Chinese documents which revealed the North Korean claim was an elaborate disinformation campaign.

During the 1950s the United States conducted a series of field tests using entomological weapons. Operation Big Itch, in 1954, was designed to test munitions loaded with uninfected fleas (*Xenopsylla cheopis*). Big Itch went awry when some of the fleas escaped into the plane and bit all three members of the air crew. In May 1955 over 300,000 yellow fever mosquitoes (*Aedes aegypti*) were dropped over parts of the U.S. state of Georgia to determine if the air-dropped mosquitoes could survive to take meals from humans. The mosquito tests were known as Operation Big

Buzz. The U.S. engaged in at least two other EW testing programmes, Operation Drop Kick and Operation May Day. A 1981 Army report outlined these tests as well as multiple cost-associated issues that occurred with EW. The report is partially declassified — some information is blacked out, including everything concerning "Drop Kick" — and included "cost per death" calculations. The cost per death, according to the report, for a vector-borne biological agent achieving a 50% mortality rate in an attack on a city was $0.29 in 1976 dollars. Such an attack was estimated to result in 625,000 deaths. The United States has also applied entomological warfare research and tactics in non-combat situations. In 1990 the U.S. funded a $6.5 million programme designed to research, breed and drop caterpillars. The caterpillars were to be dropped in Peru on coca fields as part of the American War on Drugs. As recently as 2002 U.S. entomological anti-drug efforts at Fort Detrick were focused on finding an insect vector for a virus that affects the opium poppy.

Bioterrorism

Clemson University's Regulatory and Public Service Programme listed "diseases vectored by insects" among bioterrorism scenarios considered "most likely". Because invasive species are already a problem worldwide one University of Nebraska entomologist considered it likely that the source of any sudden appearance of a new agricultural pest would be difficult, if not impossible, to determine. Lockwood considers insects a more effective means of transmitting biological agents for acts of bioterrorism than the actual agents. Insect vectors are easily gathered and their eggs easily transportable without detection. Isolating and delivering biological agents, on the other hand, is extremely challenging and hazardous. In one of the few suspected acts of entomological bioterrorism an eco-terror group known as The Breeders claimed to have released Mediterranean fruit flies (medflies) amidst an ongoing California infestation. Lockwood asserts that there is some evidence the group played a role in the event. The pest attacks a variety of crops and the state of California responded with a large-scale pesticide spraying programme. At least one source asserted that there is no doubt that an outside hand played a role in the dense 1989 infestation. The group stated in a letter to then Los Angeles Mayor Tom Bradley that their goals were twofold. They sought to cause the medfly infestation to grow out of control which, in turn, would render the ongoing Malathion spraying programme financially infeasible.

Legal Status

The Biological and Toxic Weapons Convention (BWC) of 1972 does not specifically mention insect vectors in its text. The language of the treaty, however, does cover vectors. Article I bans, "Weapons, equipment or means of delivery designed to use such agents or toxins for hostile purposes or in armed conflict." It would appear, due to the text of the BWC, that insect vectors as an aspect of entomological warfare are covered and outlawed by the Convention. The issue is less clear when warfare with uninfected insects against crops is considered.

MEDICAL ENTOMOLOGY

The discipline of medical entomology, or public health entomology, and also *veterinary entomology* is focused upon insects and arthropods that impact human health. Veterinary entomology is included in this category, because many animal diseases can "jump species" and become a human health threat, for example, bovine encephalitis (known as "mad cow disease"). Medical entomology also includes scientific research on the behaviour, ecology, and epidemiology of arthropod disease vectors, and involves a tremendous outreach to the public, including local and state officials and other stake holders in the interest of public safety. Medical Entomologists are employed by private and public universities, private industries, and federal, state, and local government agencies, including all three branches of the military - who hire medical entomologists to protect the troops from infectious diseases that can be transmitted by arthropods. Historically, during wars, more people have died due to insect-transmitted diseases, than to all the battle injuries combined. Medical entomologists are also hired by chemical companies - to help develop new pesticides which will effectively decrease insect pest populations while simultaneously protecting the health of the public. Public heath entomology has seen a huge surge in interest since 2005, due to the resurgence of the bed bug, *Cimex lectularias.*

Insects of Medical Importance

Medical entomologists work in the public health arena, dealing with insects (and other arthropods) that parasitize people, bite, sting, and/or vector disease. Personal Pests Some of which may vector pathogens: Lice, Fleas, Bedbugs, Ticks, Scabies mites

The Housefly

The housefly is a very common and cosmopolitan species which transmits diseases to man. The organisms of both amoebic and bacillary dysenteries are picked up by flies from the faeces of infected people and transferred to clean food either on the fly's hairs or by the fly vomiting during feeding. Typhoid germs may be deposited on food with the fly's faeces. The house fly cause the spread of yaws germs by carrying them from a yaws ulcer to an ordinary sore. Houseflies also transmit poliomyelitis by carrying the virus from infected faeces to food or drink. Cholera and hepatitis are sometimes fly-borne. Other diseases carried by houseflies are Salmonella, tuberculosis, anthrax, and some forms of ophthalmia. They carry over 100 pathogens and transmit some parasitic worms. The flies in poorer and lower-hygiene areas usually carry more pathogens. Some strains have become immune to most common insecticides.

Biting Insects

Mosquitoes, Biting Midges, (Sandflies), Blackflies, Horse

Flies, Stable flies. Pathogen infection transmitted by insect or other arthropod vectors. Diseases carried by insects and other arthropod vectors affect more than 700 million people every year, and are considered the most sensitive to climatic and environment conditions. (WHO)

Major

- Dengue fever - Vectors: *Aedes aegypti* (main vector) *Aedes albopictus* (minor vector) threatens -50 million people are infected by dengue annually, 25,000 die. Threatens 2.5 billion people in more than 100 countries.
- Malaria - Vectors: *Anopheles* mosquitoes - 500 million become severely ill with malaria every year and more than 1 million die.
  1. Leishmaniasis - Vectors: species in the genus *Lutzomyia* in the New World and *Phlebotomus* in the Old World. Two million people infected.
- Bubonic plague - Principal vector: *Xenopsylla cheopis* At least 100 flea species can transmit plague. Re-emerging major threat several thousand human cases per year. High pathogenicity and rapid spread.
- Sleeping sickness - Vector: Tsetse fly, not all species. Sleeping sickness threatens millions of people in 36 countries of sub-Saharan Africa (WHO)
- Typhus - Vectors: mites, fleas and body lice 16 million cases a year, resulting in 600,000 deaths annually.
- Wuchereria bancrofti - most common vectors: the mosquito species: *Culex*, *Anopheles*, *Mansonia*, and *Aedes*; affects over 120 million people.
- Yellow Fever - Principal vectors: *Aedes simpsoni*, *A. africanus*, and *A. aegypti* in Africa, species in *Haemagogus* genus in South America, and species in *Sabethes* genus in France -200,000 estimated cases of yellow fever (with 30,000 deaths) per year.

Minor

- Ross River fever - Vector: Mosquitoes, main vectors *A. vigilax*, *Aedes camptorhynchus*, and *Culex annulirostris*
  1. Barmah Forest Virus - Vector: Known vectors *Culex annulirostris*, *Ocleratus vigilax* and *O. camptorhynchus* and *Culicoides marksi*
- Kunjin encephalitis (mosquitoes)
- Murray Valley encephalitis virus (MVEV) - Major mosquito vector: *Culex annulirostris*.
- Japanese encephalitis - Several mosquito vectors, the most important being *Culex tritaeniorhynchus*.
- West Nile virus - Vectors: vary according to geographical area; in the USA *Culex pipiens* (Eastern US), *Culex tarsalis* (Midwest and West), and *Culex quinquefasciatus* (Southeast) are the main vectors.
- Lyme disease - Vectors: several species of the genus *Ixodes*
- Alkhurma virus (KFDV) - Vector: tick
- Kyasanur forest disease - Vector: *Haemaphysalis spinigera*
- Brugia timori filariasis - Primary vector: *Anopheles barbirostris*
- Babesia - Vector *Ixodes* ticks.
- Carrion's disease - Vectors: sandflies of the genus *Lutzomyia*.
- Chagas disease - Vector: assassin bugs of the subfamily *Triatominae*. The major vectors are species in the genera *Triatoma*, *Rhodnius*, and *Panstrongylus*.
- Chikungunya - Vectors: *Aedes* mosquitoes
- Human ewingii ehrlichiosis - Vector: *Amblyomma americanum*
- Human granulocytic ehrlichiosis - Vector: *Ixodes scapularis*

- Rift Valley Fever (RVF) - Vectors: fleas in the genera *Aedes* and *Culex*
- Scrub typhus - Vector: Chigger
- Loa loa filariasis - Vector: *Chrysops* sp.

LIST OF ENTOMOLOGISTS

The following is a list of entomologists, people who have studied insects.

Name	*Born*	*Died*	*Country*	*Speciality*
Lajos Abafi	1840	1909	Hungary	Lepidoptera
John Abbot	1751	1840	United States	
Camillo Acqua	1863	1936	Italy	Sericulture
Tusnamitsu Adachi	1901	1981	Japan	Coleoptera
Louis Agassiz	1807	1873	Switzerland / United States	Coleoptera
Ulisse Aldrovandi	1522	1605	Italy	
Vincent Allard	1921	1994	Belgium	Coleoptera Cetoniinae
Charles Paul Alexander	1889	1981	United States	Diptera
Wilfred Backhouse Alexander	1885	1965	United Kingdom	
Anastase Alfieri	1892	1971	Italy	Coleoptera
E.B. Alo			Nigeria	
Carlo Alonza	1881	1961	Italy	Coleoptera
William de Alwis	1842	1916	Sri Lanka	Lepidoptera
Charles Jean-Baptiste Amyot	1799	1866	France	Hemiptera
Alfredo Andreini	1870	1943	Italy	
Ross H. Arnett, Jr.	1919	1999	United States	Coleoptera
Gilbert John Arrow	1873	1948	United Kingdom	Coleoptera
William Harris Ashmead	1855	1905	United States	Hymenoptera
E.C.M. d'Assis-Fonseca			United Kingdom	Diptera
Edwin Felix Thomas Atkinson	1840	1890	Ireland	Hemiptera
William Stephen Atkinson	1820	1876	India	Lepidoptera
Charles Nicholas Aubé	1802	1869	France	Coleoptera
Jean Guillaume Audinet-Serville	1775	1858	France	Orthoptera
Jean Victoire Audouin	1797	1841	France	
Per Olof Christopher Aurivillius	1843	1928	Sweden	Coleoptera, Lepidoptera
Ernest Edward Austen	1867	1938	United Kingdom	Diptera, Hymenoptera.
Andrey Avinoff	1884	1949	Ukraine / United States	Lepidoptera
Alfred Balachowsky	1901	1983	Russia / France	Hemiptera, Coleoptera

Mary Ball	1812	1892	Ireland	Odonata, Hemiptera
Joseph Sugar Baly	1816	1890	United Kingdom	Coleoptera
John Banister (naturalist)	1650	1692	United Kingdom	
Nathan Banks	1868	1953	United States	Neuroptera, Megaloptera, Hymenoptera
Philip James Barraud	1879	1948	United Kingdom	Diptera
Eugène Barthe	1862	1945	France	Coleoptera
Agostino Bassi	1773	1856	Italy	
Henry Walter Bates	1825	1892	United Kingdom	
Marston Bates	1906	1974	United States	Diptera
Palisot de Beauvois	1752	1820	France	
Eduard Becher	1856	1886	Austria	Diptera
Theodor Becker	1840	1928	Germany	Diptera
Hans Hermann Behr	1818	1904	Germany / United States	Lepidoptera
Bryan Patrick Beirne	1918	1998	Ireland	Lepidoptera, Hymenoptera
Luigi Bellardi	1859	1889	Italy	Diptera
Joseph Charles Bequaert	1886	1982	Belgium	Diptera
May Berenbaum			United States	
Julius von Bergenstamm	1837	1896	Austria	Diptera
Georges Bernardi	1922	1999	France	Lepidoptera, Pieridae
Mario Bezzi	1868	1927	Italy	Diptera
Czesaw Bieanko	1895	1985	Poland	Lepidoptera
George Carter Bignell	1826	1910	United Kingdom	Lepidoptera and Hymenoptera (mainly Ichneumonidae)
Jacques-Marie-Frangile Bigot	1818	1893	France	Diptera
Gustaf Johan Billberg	1772	1844	Sweden	
Charles Thomas Bingham	1848	1908	Ireland	Hymenoptera
Thomas Blackburn	1844	1912	United Kingdom/ Hawaii / Australia	Coleoptera
Richard E. Blackwelder	1909	2001	United States	
Emile Blanchard	1819	1900	France	
Murray S. Blum	1929		United States	Formicidae
Carl Henrik Boheman	1796	1868	Sweden	Coleoptera
Jean Baptiste Boisduval	1799	1879	France	Lepidoptera
Willis Blatchley	1859	1940	United States	Coleoptera, Ortoptera, Hemiptera
Ignacio Bolívar	1850	1944	Spain	
Cándido Bolívar Pieltain	1897	1976	Spain	Lepidoptera
Ernst Boll	1817	1868	Germany	Lepidoptera
Barry Bolton			United Kingdom	Formicidae

Franco Andrea Bonelli	1784	1830	Italy	Coleoptera
Charles Bonnet	1720	1793	Switzerland	
George Bornemissza	1924		Australia	Coleoptera
Carl Julius Bernhard Börner	1880	1953	Germany	Collembola
Nérée Boubée	1806	1863	France	Lepidoptera, Coleoptera
Jean Bourgogne	1903	1999	France	Lepidoptera, Psychidae
Arthur I. Bourne	1886	1961	United States	Agricultural pests
Eugène Louis Bouvier	1856	1944	France	
James Henry Bowker	1822	1900	South Africa	Lepidopters
Nikolaus Joseph Brahm	1751	1812	Germany	
Karel Branèik	1842	1915	Hungary	Coleoptera
Rick Brandenburg	1955		United States	
Friedrich Moritz Brauer	1832	1904	Austria	Diptera, Neuroptera
Ferdinando Arborio Gattinara di Breme	1807	1869	Italy	Diptera, Coleoptera
Otto Vasilievich Bremer		1873	Russia	
Stephan von Breuning	1894	1983	Germany	Coleoptera Cerambycidae Lamiinae
Carl Gustav Alexander Brischke	1814	1897	Poland	Diptera, Hymenoptera
Georges Brossard	1940		Canada	
Emil von Brück	1807	1884	Germany	Coleoptera
Enrico Adelelmo Brunetti	1862	1927	United Kingdom	Diptera
William Buckler	1814	1884	United Kingdom	Lepidoptera
Jean Baptiste Lucien Buquet	1807	1889	France	Coleoptera
Hermann Burmeister	1807	1892	Germany	
John Burns (entomologist)			United States	Lepidoptera
Raymond C. Bushland	1910	1995	United States	Economic entomology
Arthur Gardiner Butler	1844	1925	United Kingdom	
Kurt Büttner	1881	1967	Germany	Heteroptera
Elsa Salazar Cade			United States	
Philip Powell Calvert	1871	1961	United States	Odonata
Giulio Camus	1847	1914	France	Economic History of Entomology
Ernest Candèze	1827	1898	Belgium	Coleoptera
Silvano Canzoneri	1941	1995	Italy	Coleoptera, Diptera
Pierre Capdeville	1908	1980	France	Lepidoptera Parnassiinae
Frank M. Carpenter	1902	1994	United States	
Herbert James Carter	1858	1940	Australia	
Thomas Lincoln Casey, Jr.	1857	1925	United States	Coleoptera
Guelfo Cavanna	1850	1920	Italy	All orders
George Charles Champion	1851	1927	United Kingdom	Coleoptera

Toussaint de Charpentier	1779	1847	Germany	
Jean-Baptiste Eugène Bellier de la Chavignerie	1844	1888	France	Lepidoptera
Lucy Evelyn Cheesman	1881	1969	United Kingdom	
Louis Alexandre Auguste Chevrolat	1799	1884	France	Coleoptera
Samuel Rickard Christophers	1873	1978	United Kingdom	Diptera
Joseph Philippe de Clairville	1742	1830	Switzerland	
Benjamin Preston Clark	1860	1939	United States	Lepidoptera
Cyril Clarke	1907	2000	United Kingdom	Lepidoptera
Eric William Classey	1916	2008	United Kingdom	Lepidoptera. Entomological publisher.
Carl Alexander Clerck	1709	1765	Sweden	Lepidoptera
Theodore Dru Alison Cockerell	1866	1948	United States	Hymenoptera
James Edward Collin	1876	1968	United Kingdom	Diptera
John Henry Comstock	1849	1931	United States	
Daniel William Coquillett	1856	1911	United States	Diptera
Matthew Cooke	1829	1887	Ireland	Economic entomology
Charles Coquerel	1822	1867	France	
Luis Ceballos y Fernández de Córdoba	1896	1967	Spain	Forest entomology, Ichneumonidae
Emilio Cornalia	1824	1882	Italy	
Achille Costa	1823	1899	Italy	
Oronzio Gabriele Costa	1787	1867	Italy	
George Crabbe	1754	1832	United Kingdom	Coleoptera
Pieter Cramer	1738	1804	Netherlands	
William Monad Crawford	1872	1941	Ireland	Lepidoptera
Henry Harpur Crewe	1828	1883	United Kingdom	Lepidoptera
George Robert Crotch	1842	1874	United States/ United Kingdom	Coleoptera
Philip Crowley	1837	1900	England	Lepidoptera
Roy Crowson	1914	1999	United Kingdom	Coleoptera
Charles Howard Curran	1894	1972	United States	Diptera
John Curtis (entomologist)	1791	1862	United Kingdom	Coleoptera
Stanis³aw Zbigniew	1963		Poland	Trichoptera
Leander Czerny	1859	1944	Czech Republic	Diptera
Anders Gustaf Dahlbom	1806	1859	Sweden	Hymenoptera
James Charles Dale	1792	1872	United Kingdom	Coleoptera
Johan Wilhelm Dalman	1787	1828	Sweden	
Franz Dannehl	1870	1947	Germany	Lepidoptera
Louis Gabriel d'Antessanty	1834	1922	France	Hemiptera
Charles Darwin	1809	1882	United Kingdom	
Thomas Reid Davys Bell	1863	1948	Ireland	Lepidoptera

Charles De Geer	1720	1778	Sweden	Coleoptera
Wilhem de Haan	1801	1855	Netherlands	
José María Hugo de la Fuente Morales	1855	1932	Spain	
Francis de Laporte de Castelnau	1810	1880	France	Coleoptera
Pierre François Marie Auguste Dejean	1780	1845	France	Coleoptera
Georges Demoulin	1919	1994	Belgium	Ephemeroptera
Robert Denno	1945	2008	United States	Lepidoptera
Michael Denis	1729	1800	Austria	Lepidoptera
Thomas Desvignes	1812	1868	England	Hymenoptera
Vincent Dethier	1915	1993	United States	
Achille Deyrolle	1813	1865	France	Coleoptera
Harry Frederick Dietz	1890	1954	United States	Economic entomology
William Lucas Distant	1845	1922	United Kingdom	
Frederick Augustus Dixey	1855	1935	United Kingdom	Lepidoptera
Theodosius Dobzhansky	1900	1975	Ukraine / United States	
Paul Dognin	1847	1931	France	Lepidoptera
William Doherty	1857	1901	United States	Lepidoptera
Carl August Dohrn	1806	1892	Germany	Coleoptera
Horace Donisthorpe	1870	1951	United Kingdom	Coleoptera, Formicidae
Charles Donovan	1863	1951	Ireland	Lepidoptera
Edward Donovan	1768	1837	Ireland	
Alcide d'Orbigny	1802	1857	France	
Edward Doubleday	1810	1849	United Kingdom	Lepidoptera
Henry Doubleday	1808	1875	United Kingdom	Lepidoptera
John William Douglas	1814	1905	United Kingdom	Lepidoptera
Herbert Druce	1846	1880	United Kingdom	Lepidoptera
Dru Drury	1725	1804	United Kingdom	
Catherine N. Duckett	1961		United States	Coleoptera
Oswald Duda	1869	1941	Germany	Diptera
Léon Jean Marie Dufour	1780	1865	France	
Caspar Erasmus Duftschmid	1767	1821	Austria	Coleoptera
Lionel Jack Dumbleton	1905	1976	New Zealand	Lepidoptera
André Marie Constant Duméril	1774	1860	France	Coleoptera
Philogène Auguste Joseph Duponchel	1774	1846	France	Lepidoptera
Jan Dzier¿on	1811	1906	Poland	Hymenoptera
Alfred Edwin Eaton	1845	1929	United Kingdom	Diptera, Ephemeroptera
Karl Eckstein	1859	1939	Germany	Lepidoptera
Frederick Wallace Edwards	1888	1940	United Kingdom	Diptera

Henry Edwards	1827	1891	United Kingdom/ United States	Lepidoptera
J. Gordon Edwards	1919	2004	United States	
William Henry Edwards	1822	1909	United States	Lepidoptera
Georg Dionysius Ehret	1708	1770	Germany	
Paul R. Ehrlich	1932		United States	Lepidoptera
Curt Eisner	1890	1981	Germany	Lepidoptera
Henry John Elwes	1846	1922	United Kingdom	Lepidoptera
Günther Enderlein	1872	1968	Germany	Diptera
Michael S. Engel	1971		United States	Zoraptera, Hymenoptera, fossils
Wilhelm Ferdinand Erichson	1809	1848	Germany	Coleoptera
Johann Friedrich von Eschscholtz	1793	1831	Estonia	
Eugenius Johann Christoph Esper	1742	1810	Germany	
Howard Ensign Evans	1919	2002	United States	Hymenoptera
W.H. Evans		1956	United Kingdom	Lepidoptera
William Frederick Evans			United Kingdom	Odonata, Orthoptera
Eduard Friedrich Eversmann	1794	1860	Germany	
Jean Henri Fabre	1823	1915	France	
Johan Christian Fabricius	1745	1808	Denmark	
Olof Immanuel von Fåhraeus	1796	1884	Sweden	Coleoptera
Graham Fairchild	1906	1994	United States	Diptera
Léon Fairmaire	1820	1906	France	Coleoptera
Carl Fredrik Fallén	1764	1830	Sweden	Diptera
Baron Cajetan von Felder	1814	1894	Austria	Lepidoptera
Rudolf Felder	1842	1871	Austria	Lepidoptera
Ephraim Porter Felt	1868	1943	United States	Diptera
Ferdinand I of Bulgaria	1861	1948	Bulgaria	Lepidoptera
Charles Henry Fernald	1838	1921	United States	Lepidoptera
Henry T. Fernald	1866	1952	United States	Economic entomology, Hymenoptera
Michael Lloyd Ferrar	1876	1931	United Kingdom	Lepidoptera
Pietro Mansueto Ferrari	1823	1893	Italy	Hemiptera
Leopold Heinrich Fischer	1817	1886	Germany	
Asa Fitch	1809	1879	United States	
Ernst Josef Fittkau	1927		Germany	Diptera
John Fleming (naturalist)	1785	1857	United Kingdom	
James Fletcher	1852	1908	Canada	
Thomas Bainbrigge Fletcher	1878	1950	United Kingdom	
E.B. Ford	1901	1988	United Kingdom	Lepidoptera

Auguste-Henri Forel	1848	1931	Switzerland	Formicidae
Arnold Förster	1810	1884	Germany	Coleoptera, Hymenoptera
Johann Reinhold Forster	1729	1798	Germany	
Margaret Fountaine	1862	1940	United Kingdom	Lepidoptera
Antoine François, comte de Fourcroy	1755	1809	France	
William Weekes Fowler	1849	1923	United Kingdom	Coleoptera
Georg Ritter von Frauenfeld	1807	1873	Austria	Diptera
Charles French (entomologist)	1842	1933	Australia	Economic entomology
Heinrich Frey	1822	1890	Switzerland	Lepidoptera
Christian Friedrich Freyer	1794	1885	Germany	Lepidoptera
Karl von Frisch	1886	1982	Austria	Hymenoptera (honey bees)
Imre Frivaldszky	1799	1870	Hungary	Lepidoptera, Coleoptera
Walter Wilson Froggatt	1858	1937	Australia	Economic entomology
Frederick William Frohawk	1861	1946	United Kingdom	Lepidoptera
Stuart W. Frost	1891	1990	United States	Diptera
Hans Fruhstorfer	1866	1922	Germany	Lepidoptera
José María Hugo de la Fuente Morales	1855	1932	Spain	Coleoptera
Johann Kaspar Füssli	1743	1786	Switzerland	
Charles Joseph Gahan	1862	1939	United Kingdom/ Ireland	Coleoptera
Burton Noble Gates	1881	1972	United States/ Canada	Apiculture, Hymenoptera
Giuseppe Géné	1800	1847	Italy	
Étienne Louis Geoffroy	1725	1810	France	Coleoptera
Ernst Friedrich Germar	1786	1853	Germany	Coleoptera
Carl Eduard Adolph Gerstaecker	1828	1895	Germany	
Paul Gervais	1816	1879	France	
Michael Thomas Gillies	1920	1999	United Kingdom	Medical entomology
Joseph-Étienne Giraud	1808	1877	France	Coleoptera, Hymenoptera
Johannes von Nepomuk Franz Xaver Gistel	1809	1873	Germany	
Eleanor Glanville	1654	1709	United Kingdom	Coleoptera
Jean Baptiste Godart	1775	1825	France	Lepidoptera
Frederick DuCane Godman	1834	1919	United Kingdom	Lepidoptera
Jan Goedart	1620	1668	Netherlands	
Maurice Emile Marie Goetghebuer	1876	1962	Belgium	Diptera
Deborah M. Gordon	1955		United States	Formicidae

Hippolyte Louis Gory	1800	1852	France	Coleoptera
William Gould	1715		United Kingdom	Formicidae
Claude Charles Goureau	1790	1879	France	
Mariano de la Paz Graëlls y de la Aguera	1809	1898	Spain	
Johann Ludwig Christian Gravenhorst	1777	1857	Germany	
Philip Graves	1876	1953	Ireland	Lepidoptera
George Robert Gray	1808	1872	United Kingdom	
Viktor Grebennikov	1927	2001	Russia	
Auguste Jean François Grenier	1814	1890	France	Coleoptera
Thomas de Grey, 6th Baron Walsingham	1843	1919	United Kingdom	Lepidoptera
David Grimaldi	1957		United States	Diptera
Augustus Radcliffe Grote	1841	1903	United Kingdom	Lepidoptera
Grigory Grumm-Grzhimaylo	1860	1936	Russia	Lepidoptera
Achille Guenée	1809	1880	France	Lepidoptera
Félix Édouard Guérin-Méneville	1799	1874	France	
Guido of Pisa		1169	Italy	
Leonard Gyllenhaal	1752	1840	Sweden	Coleoptera
Hermann August Hagen	1817	1893	Germany	Neuroptera, Odonata
James Nathaniel Halbert	1871	1948	Ireland	
Alexander Henry Haliday	1807	1870	Ireland	Coleoptera, Diptera
George Hampson	1860	1936	United Kingdom	Lepidoptera
D. Elmo Hardy	1914	2002	United States	Diptera
Moses Harris	1731	1785	United Kingdom	
Thaddeus William Harris	1795	1856	United States	
Michael Hassell			United Kingdom	
Hermann Haupt	1873	1959	Germany	Hymenoptera
Adrian Hardy Haworth	1767	1833	United Kingdom	
John Heath (entomologist)			United Kingdom	
Johann Christian Ludwig Hellwig	1743	1831	Germany	
Friedrich Georg Hendel	1874	1936	Austria	Diptera
Willi Hennig	1913	1976	Germany	Diptera
Johann Friedrich Wilhelm Herbst	1743	1807	Germany	Coleoptera
Claude Herbulot Geometridae	1908	2006	France	Lepidoptera
Hans Rudolf Herren	1947		Switzerland	Economic entomology
Gottlieb August Wilhelm Herrich-Schäffer	1799	1874	Germany	
William Chapman Hewitson	1806	1878	United Kingdom	Coleoptera, Lepidoptera

H. E. Hinton	1912	1977	United Kingdom	Coleoptera
Jacob Hoefnagel	1575	1630	Netherlands	
Johann Centurius Hoffmannsegg	1766	1849	Germany	
William Jacob Holland	1848	1932	United States	Lepidoptera
Bert Hölldobler	1936		Germany	Formicidae
August Emil Holmgren	1829	1888	Sweden	Hymenoptera
Frederick William Hope	1797	1862	United Kingdom	Coleoptera
George Henry Horn	1840	1897	United States	Coleoptera
Leland Ossian Howard	1857	1950	United States	Economic entomology
Jacob Hübner	1761	1826	Germany	Lepidoptera
George Vernon Hudson	1867	1946	United Kingdom	
Johann Siegfried Hufnagel	1724	1795	Germany	Lepidoptera
Alphonse Hustache	1872	1942	France	Coleoptera, Curculionidae
Johann Karl Wilhelm Illiger	1775	1813	Germany	Coleoptera
Augustus Daniel Imms	1880	1949	United Kingdom	
Carl Gustav Jablonsky	1756	1787	Germany	Coleoptera
Jean-Charles Jacobs	1821	1907	Belgium	Diptera, Hymenoptera
Georgij Georgiewitsch Jacobson	1871	1926	Russia	
Tadeusz Jaczewski	1899	1974	Russia	Hemiptera
René Jeannel	1879	1965	France	Coleoptera
Anders Christian Jensen-Haarup	1863	1934	Denmark	Hymenoptera
Thomas C. Jerdon	1811	1872	United Kingdom	Formicidae
William Frederick Johnson	1852	1934	Ireland	
James John Joicey	1871	1932	United Kingdom	Lepidoptera
Karl Jordan	1861	1959	Germany	Siphonaptera, Coleoptera, Lepidoptera
Norman H. Joy	1874	1953	United Kingdom	Coleoptera
Louis Jurine	1751	1819	Switzerland	Coleoptera, Diptera
Vlasta Kálalová	1896	1971	Czech Republic	
Johann Heinrich Kaltenbach	1807	1876	Germany	Economic entomology
William Francis de Vismes Kane	1840	1918	Ireland	Lepidoptera
David Keilin	1887	1963	United Kingdom	
Kálmán Kertész	1867	1922	Hungary	Diptera
Bernard Kettlewell	1907	1979	United Kingdom	Lepidoptera
Jean-Jacques Kieffer	1857	1925	France	Diptera, Hymenoptera
Ernest August Hellmuth von Kiesenwetter	1820	1880	Germany	Coleoptera
Alfred Kinsey	1894	1956	United States	Hymenoptera
William Kirby (entomologist)	1759	1850	United Kingdom	
William Forsell Kirby	1844	1912	United Kingdom/ Ireland	Lepidoptera

Ruud Kleinpaste	1952		Netherlands / New Zealand	
Johann Christoph Friedrich Klug	1775	1856	Germany	Lepidoptera
Edward F. Knipling	1909	2000	United States	Economic entomology
Carl Ludwig Koch	1778	1857	Germany	
Vincenz Kollar	1797	1860	Germany	Diptera
Friedrich Wilhelm Konow	1842	1908	Germany	Hymenoptera Tenthredinidae
Ferdinand Kowarz	1838	1914	Austria	Diptera
Ernst Gustav Kraatz	1831	1909	Germany	Coleoptera
Otto Kröber	1882	1969	Germany	Diptera
Johann Gottlieb Kugelann	1753	1815	Germany	Coleoptera
Joseph Alexandre Laboulbène	1825	1898	France	
Jean Théodore Lacordaire	1801	1870	Belgium	Coleoptera
Jean-Pierre Lacroix	1938	1989	France	Coleoptera Lucanidae
Johann Nepomuk von Laicharting	1754	1797	Austria	Coleoptera
Eric Laithwaite	1921	1997	United Kingdom	Lepidoptera
Sven Lampa	1839	1914	Sweden	Lepidoptera
Pierre André Latreille	1762	1833	France	
John Eatton Le Conte	1784	1860	United States	
William Elford Leach	1790	1836	United Kingdom	
John Lawrence LeConte	1825	1883	United States	Coleoptera
Julius Lederer (entomologist)	1821	1870	Austria	Lepidoptera
Alexandre Louis Lefèbvre de Cérisy	1798	1867	France	
Claude Lemaire	1921	2004	France	Lepidoptera Saturniidae
Amédée Louis Michel Lepeletier de Saint-Fargeau	1780	1845	France	Hymenoptera
Erwin Lindner	1888	1988	Germany	Diptera
Carl Linnaeus	1707	1778	Sweden	
Melchior de Lisle	1908	1977	France	Coleoptera Lucanidae
Hermann Loew	1807	1879	Germany	Diptera
Cynthia Longfield	1896	1991	Ireland	Odonata
Hippolyte Lucas	1814	1899	France	
William Lundbeck	1863	1941	Denmark	Diptera
Adolfo Lutz	1855	1940	Brazil	Medical entomology
Robert Mac Lachlan	1837	1904	United Kingdom	Neuroptera
Alexander Macleay	1767	1848	United Kingdom	
William Sharp Macleay	1792	1865	United Kingdom	
Justin Pierre Marie Macquart	1776	1855	France	Diptera
Michael Majerus	1954		United Kingdom	Lepidoptera
René Malaise	1892	1978	Sweden	

John Russell Malloch	1875	1963	United Kingdom	Diptera
Mahadeva Subramania Mani	1908	2003	India	
Josef Johann Mann	1804	1889	Czech Republic	Lepidoptera
Carl Gustaf von Mannerheim	1797	1854	Finland	Coleoptera
Patrick Manson	1844	1922	Scotland	Medical entomology
Oldøich Marek	1911	1986	Czech Republic	Coleoptera
Charles Lester Marlatt	1863	1954	United States	Economic entomology
Sylvain Auguste de Marseul	1812	1890	France	Coleoptera
George Frederick Leycester Marshall	1843	1934	United Kingdom	Lepidoptera
Thomas Ansell Marshall	1827	1903	United Kingdom	Hymenoptera
Thomas Marsham		1819	United Kingdom	Coleoptera
Fermín Martín Piera	1954	2001	Spain	Coleoptera
Andrey Vasilyevich Martynov	1879	1938	Russia	Trichoptera
William Miles Maskell	1839	1898	New Zealand	
Wayne Masterson	1959	1991	United Kingdom	Medical entomology
Shonen Matsumura	1872	1960	Japan	
Harold Maxwell-Lefroy	1877	1926	United Kingdom	
Henry Christopher McCook	1837	1911	United States	Formicidae
Albert Stewart Meek	1871	1943	United Kingdom	
Jean Pierre Mégnin	1828	1905	France	Forensic entomology
Johann Wilhelm Meigen	1764	1845	Germany	Diptera
Axel Leonard Melander	1878	1962	United States	Diptera, Hymenoptera
Frederick Valentine Melsheimer	1749	1814	United States	Coleoptera
Ernest Menault	1830	1903	France	Agricultural pests
Édouard Ménétries	1802	1861	France / Russia	Lepidoptera, Coleoptera
Maria Sibylla Merian	1647	1717	Germany	
Gustav Mayr	1830	1908	Austria	Hymenoptera
Edward Meyrick	1854	1938	United Kingdom	Lepidoptera
Charles Duncan Michener	1918		United States	Hymenoptera
Josef Mik	1839	1900	Czech Republic	Diptera
Pierre Millière	1811	1887	France	Lepidoptera
Adolphe Philippe Millot	1857	1921	France	Lepidoptera
Adolph Modéer	1738	1799	Sweden	
Frederic Moore	1830	1907	United Kingdom	Lepidoptera
Clodoveo Carrión Mora	1883	1957	Ecuador	
Claude Morley	1874	1951	United Kingdom	Hymenoptera
Derek Wragge Morley	1920	1969	United Kingdom	Formicidae
Roger Morse	1927	2000	United States	Hymenoptera
Kenneth Morton			United Kingdom	Odonata, Neuroptera
Victor Motschulsky	1810	1871	Russia	Coleoptera

Thomas Muffet	1552	1604	United Kingdom	
Johann Carl Megerle von Mühlfeld	1765	1840	Austria	Coleoptera
Otto Friedrich Müller	1730	1784	Denmark	
Étienne Mulsant	1797	1880	France	
Vladimir Nabokov Lepidoptera	1899	1977	Russia / United States	
Longinos Navás	1858	1938	Spain	Neuropteroidea
Christian Gottfried Daniel Nees von Esenbeck	1776	1858	Germany	
Nikolaus Poda von Neuhaus	1723	1798	Austria	
Edward Newman	1801	1876	United Kingdom	
L. Hugh Newman			United Kingdom	Lepidoptera
George Newport	1803	1854	United Kingdom	
Lionel de Nicéville	1852	1901	United Kingdom	Lepidoptera
Ebbe Schmidt Nielsen	1950	2001	Denmark	
John Nietner		1874	Germany	Coleoptera
Guido Nonveiller	1913	2002	Croatia	Coleoptera
Kenneth Richard Norris	1914	2003	Australia	
Charles Oberthür	1845	1924	France	Lepidoptera
Georg Hermann Alexander Ochs	1887	1971	Germany	Coleoptera
Ferdinand Ochsenheimer	1767	1822	Germany	Lepidoptera
Thomas R. Odhiambo	1931	2003	Kenya	Medical entomology
Lawrence Ogilvie	1898	1980	Bermuda & United Kingdom	Insects of Bermuda
Harold Oldroyd	1914	1978	United Kingdom	Diptera
Guillaume-Antoine Olivier	1756	1814	France	
Eugene O'Mahoney	1899	1951	Ireland	Coleoptera, Mallophaga, Siphonaptera
Paul W. Oman	1908		United States	Hemiptera
Eleanor Anne Ormerod	1828	1901	United Kingdom	
Karl Robert Osten-Sacken	1828	1906	Russia	Diptera
Alpheus Spring Packard	1839	1905	United States	
Louis Pandellé	1824	1905	France	Coleoptera
Georg Wolfgang Franz Panzer	1755	1829	Germany	Coleoptera
Francis Polkinghorne Pascoe	1813	1893	United Kingdom	Coleoptera
Carlo Passerini	1793	1857	Italy	Coleoptera
Giovanni Passerini	1816	1893	Italy	Hemiptera
Gustaf von Paykull	1757	1826	Sweden	Coleoptera
Titian Peale	1799	1885	United States	Lepidoptera
William Dandridge Peck	1763	1822	United States	
Edward Pelham Clinton, 10th Duke of Newcastle under-Lyne	1920	1988	United Kingdom	Lepidoptera

Achille Rémy Percheron	1797	1869	France	
Alexander Petrunkevitch	1875	1964	Russia	
Maurice Pic	1866	1957	France	Coleoptera, Cerambycide
François Jules Pictet de la Rive	1809	1872	Switzerland	
Elliot Pinhey	1910	2000	United Kingdom	Lepidoptera, Odonata
Odorado Pirazzoli	1815	1884	Italy	Coleoptera
George Poinar, Jr.			United States	
Robert Michael Pyle	1947		United States	Lepidoptera
Oktawiusz Bourmeister-Radoszkowski	1820	1895	Russia	Hymenoptera
Constantine Samuel Rafinesque	1783	1840	France	
Jules Pierre Rambur	1801	1870	France	
Maynard Jack Ramsay	1914	2005	United States	Economic entomology
Alex Rasnitsyn			Russia	Palaeoentomology
Julius Theodor Christian Ratzeburg	1801	1871	Germany	
Pierre Réal	1922	2009	France	Lepidoptera, Morphidae, Pieridae, Tortricidae
René Antoine Ferchault de Réaumur	1683	1757	France	
Ludwig Redtenbacher	1814	1876	Austria	Coleoptera
Louis Jérôme Reiche	1799	1890	Netherlands / France	Coleoptera
George Morrison Reid Henry	1891	1983	Sri Lanka	Orthoptera
Edmund Reitter	1845	1920	Germany	Coleoptera
James Rennie Lepidoptera	1787	1867	Scotland	General entomology,
Anders Jahan Retzius	1742	1821	Sweden	
Enzio Rafael Reuter	1867	1951	Finland	Lepidoptera
Max Paul Riedel	1870	1941	Germany	Diptera
Charles Valentine Riley	1843	1895	United States	Economic entomology
Robert Henry Fernando Rippon	1816	1917	United Kingdom	Lepidoptera
Jean-Baptiste Robineau-Desvoidy	1799	1857	France	Diptera
Harold E. Robinson	1932		United States	Diptera
Johann Jacob Roemer	1763	1819	Switzerland	
Alois Friedrich Rogenhofer	1831	1897	Austria	Lepidoptera
Boris Borisovitsch Rohdendorf	1904	1977	Russia	Diptera
Hermann Rolle	1864	1929	Germany	Lepidoptera, Insect dealer
Per Abraham Roman	1872	1943	Sweden	Hymenoptera
Camillo Róndani	1808	1879	Italy	Diptera
August Johann Rösel von Rosenhof	1705	1759	Germany	

Pietro Rossi	1738	1804	Italy	Odonata, Lepidoptera
Josef Emanuel Fischer von Röslerstamm	1787	1866	Czech Republic	Lepidoptera
Charles Rothschild	1877	1923	United Kingdom	Siphonaptera
Miriam Louisa Rothschild	1908	2005	United Kingdom	Siphonaptera
Walter Rothschild, 2nd Baron Rothschild	1868	1937	United Kingdom	Lepidoptera
S. A. von Rottemburg			Germany	Lepidoptera
G T. Rudd			United Kingdom	Coleoptera
Gaston Ruter	1898	1979	France	Coleoptera Cetoniidae
Osbert Salvin	1835	1898	United Kingdom	
George Samouelle	1790	1846	United Kingdom	Lepidoptera
Felix Santschi	1872	1940	Switzerland	Formicidae
Felicien Henry Caignart de Saulcy	1832	1912	France	Coleoptera
William Wilson Saunders	1809	1879	United Kingdom	Lepidoptera, Hymenoptera
Henri de Saussure	1829	1905	Switzerland	Hymenoptera, Orthoptera
Thomas Say	1787	1834	United States	
Johann Gottlieb Schaller	1734	1814	Germany	Lepidoptera
William Schaus	1858	1942	United States	Lepidoptera
Johann Rudolph Schellenberg	1740	1806	Switzerland	Coleoptera, Diptera
Ignaz Schiffermüller	1727	1806	Austria	Lepidoptera
Ignaz Rudolph Schiner	1813	1873	Austria	Diptera
Jørgen Matthias Christian Schiødte	1815	1884	Denmark	
Dietrich von Schlechtendal	1834	1916	Germany	Hymenoptera, Cynipidae
Justin O. Schmidt			United States	
Hermann Schmitz	1878	1960	Germany	Diptera, Phoridae.
Carl Johan Schönherr	1772	1848	Sweden	Coleoptera
Franz Paula von Schrank	1747	1835	Germany	
Johann Christian Daniel von Schreber	1739	1810	Germany	
Arnold Schultze	1875	1948	Germany	Lepidoptera
Theodor Emil Schummel	1786	1848	Germany	Diptera
Giovanni Antonio Scopoli	1723	1788	Italy	Coleoptera
Alexander Walker Scott	1800	1883	Australia	
John Scott (entomologist)	1823	1888	United Kingdom	
Samuel Hubbard Scudder	1837	1911	United States	
Boris Nikolayevich Schwanwitsch	1889	[illegible]	Russia	Lepidoptera

Wolfgang Schwenke	1921	2006	Germany	Forest entomology, Ichneumonidae
Eugène Séguy	1890	1985	France	Diptera
Adalbert Seitz	1860	1938	Germany	Lepidoptera
Edmond de Sélys Longchamps	1813	1900	Belgium	Odonata
Andrey Semyonov-Tyan-Shansky	1866	1942	Russia	Coleoptera
Ronald A. Senior-White	1891	1954	United Kingdom	Diptera
David Sharp (entomologist)	1840	1922	United Kingdom	Coleoptera
Herbert Kenneth Airy Shaw	1902	1985	United Kingdom	
Robert Walter Campbell Shelford	1872	1912	United Kingdom	
Philip Sheppard	1921	1976	United Kingdom	Lepidoptera
Takashi Shirozu	1917	2004	Japan	Lepidoptera
Frédéric Jules Sichel	1802	1868	France	Hymenoptera
Victor Antoine Signoret	1816	1889	France	Hemiptera
Sydney Skaife	1889	1976	South Africa	Hymenoptera
Frederick Smith (entomologist)	1805	1879	United Kingdom	Hymenoptera
James Edward Smith	1759	1828	United Kingdom	
Ray F. Smith	1919	1999	United States	Economic entomology
Sidney Irving Smith	1843	1926	United States	
Roy Snelling	1934	2008	United States	Hymenoptera
Robert Evans Snodgrass	1875	1962	United States	
Árpád Soós	1912	1991	Hungary	Diptera
Paul Gustav Eduard Speiser	1877	1945	Germany	Diptera
William Spence	1783	1860	United Kingdom	
William Blundell Spence	1813	1900	United Kingdom	
Andrew Spielman	1930	2006	United States	Medical entomology
Maximilian Spinola	1780	1857	France	Coleoptera, Hymenoptera, Hemiptera
Rasmus Carl Staeger	1800	1875	Denmark	Diptera
Henry Tibbats Stainton	1822	1892	United Kingdom	Lepidoptera
Carl Stål	1833	1878	Sweden	Hemiptera
Hermann Friedrich Stannius	1808	1883	France	Diptera
Otto Staudinger	1830	1900	Germany	Lepidoptera
Pietro Stefanelli	1835	1919	Italy	Odonata, Lepidoptera
Paul Stein (entomologist)	1852	1921	Germany	Diptera
Samuel Friedrich Stein	1818	1885	Germany	Diptera
James Francis Stephens	1792	1852	United Kingdom	Coleoptera
Christian von Steven	1781	1863	Russia	Lepidoptera
Hans Ferdinand Emil Julius Stichel	1862	1936	Germany	Lepidoptera

Caspar Stoll		1795	Netherlands	
Embrik Strand	1876	1947	Norway	
Ferdinand Heinrich Hermann Strecker	1836	1901	United States	Lepidoptera
Gabriel Strobl	1846	1925	Austria	Diptera
Henry Frederick Strohecker	1905	1988	United States	Coleoptera
Johann Heinrich Sulzer	1735	1813	Switzerland	
Surcouf	1873	1934	France	
William Swainson	1789	1855	United Kingdom	
Jan Swammerdam	1637	1680	Netherlands	
Charles Swinhoe	1838	1923	United Kingdom	Lepidoptera
Robert Swinhoe	1836	1877	United Kingdom	
Datuk Rahman Anwar Syed	1932	2009	Pakistan	Economic entomology
Zoltán Szilády	1878	1947	Hungary	Diptera
Stephen Taber III	1924	2008	United States	Hymenoptera
Ernst Ludwig Taschenberg	1818	1897	Germany	
Robert Templeton	1802	1892	Ireland	
Gilles Terral	1943	1998	France	Lepidoptera, Saturniidae
Oskar Theodor	1898	1987	Israel	Diptera
Cyrus Thomas	1825	1910	United States	
Michael C. Thomas			United States	Coleoptera
Carl Gustaf Thomson	1824	1899	Sweden	Coleoptera, Hymenoptera
Carl Peter Thunberg	1743	1828	Sweden	Coleoptera
Robert John Tillyard	1881	1937	Australia	Odonata, Plecoptera, Neuroptera
Philip Hunter Timberlake	1883	1981	United States	Hymenoptera
Norman Tindale	1900	1993	Australia	
Friedrich F. Tippmann	1894	1974	Hungary	Coleoptera
Maeda Toshiyasu	1799	1859	Japan	All orders
Hervé de Toulgoët	1911	2009	France	Lepidoptera Arctiidae
Adolfo Targioni Tozzetti	1823	1902	Italy	Hemiptera
Roland Trimen	1840	1916	United Kingdom/ South Africa	
Alessandro Trotter	1874	1967	Italy	Hymenoptera and Diptera
Charles Henry Turner	1867	1923	United States	All orders
J.W. Tutt	1858	1911	United Kingdom	Lepidoptera
Iwasaki Tsunemasa	1786	1842	Japan	All higher orders
Peter Twinn	1916	2004	United Kingdom	Coleoptera
Pierre Léonard Vander Linden	1797	1831	Belgium	
Ruggero Verity	1883	1959	Italy	Lepidoptera

George Henry Verrall	1855	1911	United Kingdom	Diptera
Enrico Verson	1845	1927	Italy	Sericulture
Carlo Vidano	1923	1989	Italy	Auchenorrhyca, Apiculture
Pierre Viette	1921		France	Lepidoptera
Joseph Villeneuve de Janti	1868	1944	France	Diptera
Sigbert Wagener	1919	2004	Germany	Lepidoptera
David L. Wagner	1956		United States	Lepidoptera
Johan August Wahlberg	1810	1856	Sweden	Coleoptera, Lepidoptera
Peter Fredrik Wahlberg	1800	1877	Sweden	
Charles Athanase Walckenaer	1771	1852	France	
Johann Fischer von Waldheim	1771	1853	Germany	
James John Walker (entomologist)	1851	1939	United Kingdom	
Francis Walker (entomologist)	1809	1874	United Kingdom	
Hans Daniel Johan Wallengren	1823	1894	Sweden	
Benjamin Dann Walsh	1808	1869	United States	
Joseph Waltl	1805	1888	Germany	Coleoptera
Doug Waterhouse	1916	2000	Australia	
Frederick George Waterhouse	1815	1898	Australia	
George Robert Waterhouse	1810	1888	United Kingdom	
John Jenner Weir	1822	1894	United Kingdom	Lepidoptera
Torkel Weis-Fogh	1922	1975	Denmark	Orthoptera
Constantin Wesmael	1798	1872	Belgium	Hymenoptera
Mary Jane West-Eberhard			United States	Hymenoptera
Bernt Wilhelm Westermann	1781	1868	Denmark	
John Obadiah Westwood	1805	1893	United Kingdom	Formicidae
William Morton Wheeler	1865	1937	United States	Formicidae
Adam White (zoologist)	1817	1879	United Kingdom	
Francis Buchanan White	1842	1894	United Kingdom	
Christian Rudolph Wilhelm Wiedemann	1770	1840	Germany	Diptera
Vincent Wigglesworth	1899	1994	United Kingdom	
Benjamin Wilkes			United Kingdom	Lepidoptera
Carrington Bonsor Williams	1889	1981	United Kingdom	Economic entomology, migration, ecology
E. O. Wilson	1929		United States	Formicidae
Johannes Winnertz	1800	1896	Germany	Diptera
Wilhelm von Winthem	1799	1847	Germany	Diptera, Hymenoptera
Heinrich Wolf	1924	2004	Germany	Hymenoptera
Johann Friedrich Wolff	1778	1806	Germany	Hymenoptera
Thomas Vernon Wollaston	1822	1878	United Kingdom	Coleoptera
Charles E. Woodworth	1897	1966	United States	Economic entomology

Charles W. Woodworth	1865	1940	United States	Economic entomology
Thomas Workman	1844	1900	Ireland	Lepidoptera
Philogène Auguste Galilée Wytsman	1866	1925	Belgium	
Philipp Christoph Zeller	1808	1883	Germany	
Johan Wilhelm Zetterstedt	1785	1875	Sweden	Diptera, Hymenoptera
Daniel Ziegler	1804	1876	United States	
Johann Leopold Theodor Friedrich Zincken	1770	1856	Germany	
Fritz Konrad Ernst Zumpt	1908	1985	Germany	Diptera

REFERENCES

Baker, Eric (December 25, 2008). "'Maggot bombs' and malaria", *Casper Star-Tribune*, via the *Laramie Boomerang*, February 27, 2006, accessed.

Baumann, Peter (December 23, 2008). "Warfare gets the creepy-crawlies", *Laramie Boomerang*, October 18, 2008, accessed.

Croddy, Eric and Wirtz, James J. (2005).*Weapons of Mass Destruction: An Encyclopedia of Worldwide Policy, Technology, and History*, (Google Books), ABC-CLIO, p. 304.

Desowitz, R.S. (1991). *The malaria capers*. Norton and Co., New York, NY.

Lockwood, Jeffrey A. (December 23, 2008). "Bug Bomb", *Boston Globe*, October 21, 2007, accessed.

McNeil, W.H. (1976). *Plagues and people*. Anchor Press, Doubleday, Garden City, NY.

Mullen, G.L., and Durden, L.A., eds. (2002). *Medical and Veterinary Entomology*, Academic Press, NY

Peterson, R.K.D. (Dec. 2008). "The Role of Insects as a Biological Weapon", Department of Entomology, *Montana State University*, notes based on seminar, 1990, accessed.

Phillips, Michele R. (15 Apr. 2008). "Other insects of forensic importance." Forensic Entomology Evidence Training.

Service, M. (2008). *Medical Entomology for Students* 4th Edition Cambridge University Press.

Wheelis, Mark (2006). et al. *Deadly Cultures: Biological Weapons Since 1945*, (Google Books), Harvard University Press, pp. 84-90.

Zinsser, H. (1934). *Rats, lice, and history*. Little, Brown, and Co., New York, NY.

Bibliography

Alex Rasnitsyn & D.L.J. Quicke (2002). *History of Insects*. Kluwer Academic Publishers.

Alley N.F., Frakes L.A. (2003). "First known Cretaceous glaciation: Livingston Tillite, South Australia". *Australian Journal of Earth Science* 50: 134–150.

Baker, R.R. (1987). Integrated use of moon and magnetic compasses by the heart-and-dart moth, *Agrotis exclamationis*. Animal Behaviour 35, 94–101.

Benecke M. (2001). "A brief history of forensic entomology". *Forensic. Sci. Int.* 120 (1–2): 2–14.

Benisch, Christoph (2010). "Phylogeny of the beetles". *The beetle fauna of Germany*. Kerbtier.

Cabera, Brian J. (2007-10). "drugstore beetle- Stegobium paniceum". *University of Florida*. The University of Florida.

Campobasso, CP; Di Vella, G; Introna, F (2001). "Factors affecting decomposition and Diptera colonization". *Forensic science international* 120 (1–2): 18–27.

Capinera, John L. (Editor). (2008). *Encyclopedia of Entomology*, (2nd ed). Springer Reference.

Catts, E. P.; Goff, M.L. (January 1992). "Forensic Entomology in Criminal Investigations". *Annual Review of Entomology* 37: 253–272.

Chapman R.F. (1998). The Insects Structure and Function 4th ed. Cambridge University Press 520-522.

Chapman, A.D. (2006). *Numbers of living species in Australia and the World*. Canberra: Australian Biological Resources Study, p. 60.

Chapman, R.F. (1998). *The Insects : Structure and Function*. 4th Edition. Cambridge University Press.

Christopher O'Toole (2002). *Firefly Encyclopedia of Insects and Spiders*. Toronto: Firefly Books.

Duncan, Carl D. (1939). *A Contribution to The Biology of North American Vespine Wasps* (1 ed.). Stanford: Stanford University Press, pp. 24–29.

Elias, S.A. (2000). "Late Pleistocene Climates of Beringia, Based on Analysis of Fossil Beetles". *Quaternary Research* 53 (2): 229–235.

Ellington, CP (1999). "The novel aerodynamics of insect flight: Applications to micro-air vehicles" (PDF). *The Journal of experimental biology* 202 (Pt 23): 3439–48.

Elzinga, R.J. (2004). *Fundamentals of Entomology* (6th Ed) Pearson / Prentice Hall, New Jersey USA

H.V. Hoell, J. T. Doyen & A.H. Purcell (1998). *Introduction to Insect Biology and Diversity* (2nd ed.). Oxford University Press. pp. 407–409.

John R. Meyer (March 8, 2005). "Trichoptera". *ENT 425 - General Entomology*. North Carolina State University.

Klotzbach H, Krettek R. *et al.* (2004). "The history of forensic entomology in German-speaking countries". *Forensic. Sci. Int.* 144 (2–3): 259–263.

Mishra AK, Singh N, Sharma V.P. (1995). "Use of neem oil as a mosquito repellent in tribal villages of mandla district, madhya pradesh", Indian J Malariol, Sep. 32(3): 99-103 *Pubmed*

Möller, R. (2002). (in German). *A Biorobotics Approach to the Study of Insect Visual Homing Strategies*, p. 11.

North, David T.; Holt, Gerald (1968). Inherited Sterility in Progeny of Irradiated Male Cabbage.

Oliver Béthoux (2009). "The earliest beetle identified". *Journal of Paleontology* 83 (6): 931–937.

Payne, J.A. (1965). "A summer carrion study of the baby pig sus scrofa Linnaeus". *Ecology* 46 (5): 511–523.

Perez, S.M., Taylor, O.R. and Jander, R. (1999). The effect of a strong magnetic field on monarch butterfly (*Danaus plexippus*) migratory behavior. Naturwissenschaften 86, 140–143.

Pierre Jolivet, João Vasconcellos Neto & Philip Weinstein (1990). "Cycloalexy: a new concept in the larval defense of insects". *Insecta Mundi* 4 (1–4): 133–141.

Platzer, Max F. (2005). Department Chairman, "Aerodynamics and Aeroelasticity: Flapping-Wing Propulsion." Last revised.

Rékási J et al. (1997). Patterns in the distribution of avian lice (Phthiraptera: Amblycera, Ischnocera). *Journal of Avian Biology,* 28, 150–156.

Rodriguez W.C., Bass W.M. (1983). "Insect activity and its relationship to decay rates of human cadavers in East Tennessee". *Journal of Forensic Sciences* 28 (2): 423–432.

S. Tz'u., B.E. Mc Knight (1981). The Washing Away Of Wrongs, Center for Chinese Studies The University of Michigan, Pages 1-34.

Taylor, L.R. (1974). Insect migration, flight periodicity and the. Boundary layer. J. Anim. EcoL 43: 225-238.

Tracey, J et al.; Wilson, RI; Laurent, G; Benzer, S (18 April 2003). "Painless, a Drosophila gene essential for nociception". *Cell* 113 (2): 261–273.

Tracey, J., W. Daniel, R. I. Wilson, G. Laurent, and S. Benzer (2003). *Painless*, a *Drosophila* gene essential for nociception. Cell 113: 261-273.

Triplehorn, C.A. and Johnson, N.F. (2005). *Borror and DeLong's Introduction to the Study of Insects* (7th Ed). Brooks / Thomson Cole USA.

Tsutsui, Neil D.; Suarez Andrew V. (2003). "The Colony

Structure and Population Biology of Invasive Ants". *Conservation Biology* 17 (1): 48–58.

W.S. Robinson, R. Nowogrodski & R.A. Morse (1989). "Pollination parameters". *Gleanings in Bee Culture* 117: 148–152.

Weiss, M.L., & Mann, A.E. (1985). *Human Biology and Behaviour: An Anthropological Perspective.*. Boston: Little Brown & Co.

Whitney, Stephen R. & Sandelin, R. (2004). *Field Guide to the Cascades & Olympics*. The Mountaineers Books, p. 317.

Index

A

Achille Costa, 263
Adolfo Targioni Tozzetti, 75
Adrian Hardy Haworth, 267
Alessandro Trotter, 275
Alexander Walter Scott, 29
Alfred Russel Wallace, 29
Alfredo Andreini, 260
Alpheus Spring Packard, 271
Anastase Alfieri, 260
Anastrepha ludens, 194
Anders Gustaf Dahlbom, 10
Andrey Avinoff, 260
Antonio Berlese, 75
Apiology, 3
Archaeognatha, 150, 206
Archodonata, 214
Augustus Daniel Imms, 37

B

Benjamin Wilkes, 277
Bioterrorism, 255
Blaptica dubia, 99
Blattaria, 135, 215

C

Caloneurodea, 135, 215
Camillo Acqua, 75, 260
Camillo Róndani, 75
Carabidae, 143
Carboniferous, 134
Carcass beetles, 59
Carl Alexander Clerck, 263
Carl August Dohrn, 10
Carl Fredrik Fallén, 12
Carl Gustaf Mannerheim, 21
Caspar Erasmus, 9
Ceratitis capitata, 194
Charles Henry Turner, 275
Charles Nicolle, 39
Charles Paul Alexander, 39
Charles W. Woodworth, 76, 277
Chrysomeloidea, 141
Coleoptera, 104
Coleopterology, 3
Coxoplectoptera, 207
Cranston, 164, 175
CRC, 135
Cryoprotectants, 127
Cydia pomonella, 191
Cyril Clarke, 262

D

David L. Wagner, 276
DDT, 113

Dermaptera, 215
DGC, 178
Diaphanopterodea, 215
Dicondylia, 112
Dipterology, 3
DNA, 114, 136
Drugstore Beetle, 78
Dytiscidae, 143

E

Edward Meyrick, 270
Elateridae, 143
Elzinga, 175
Embioptera, 215
Embrik Strand, 274
Endocrine System, 169
Enrico Verson, 31, 75
Entomological Evidence Collection, 237
Entomology, 1
Entomophagous Parasite, 223
Entomophagy, 116, 216, 218
Ephemeroptera, 160, 214
Ephemeropterans, 135
Erwin Lindner, 40
Eusociality, 178

F

FDA, 86, 89
Regulations, 85
Felix Santschi, 273
Flour Beetles, 77
Forensic Entomology, 50, 65
Francesco Redi, 52
Friedrich Zincken, 277
Fuente Morales, 263

G

G.B. Reed, 251
Gabriel Strobl, 274
Gaston Ruter, 273
George Perry, 12
Gil Grissom, 2
Global climate conditions, 132
Glosselytrodea, 216
Gordon Williams, 2
Grylloblattodea, 215
Guelfo Cavanna, 262
Gullan, 164, 175
Gustaf Johan Billberg, 15
Gustaf von Paykull, 12
Gustavo Leonardi, 75

H

Ham beetles, 59
Harold Oldroyd, 271
Heinrich Frey, 266
Hemipterology, 3
Henri de Saussure, 273
Henri Marie, 13
Hermann Reinhard, 53
Hibernating Insects, 127
Hymenoptera, 3, 104
Hymenopterans, 111
Hyperparasitism, 224

I

IAEA, 198
Ice Nucleators, 126
IFTA, 229
Insect,
Biodiversity, 227

Ecology, 116
Phylogeny, 129
Intracellular Freezing, 127
Isoptera, 215

J

J.W. Tutt, 275
Jacob Hübner, 9, 13
James Francis Stephens, 274
James P. Hogan, 2
Jan Goedart, 266
Jean Baptiste, 7
Jean Baptiste Boisduval, 12
Jean Pierre Mégnin, 54
Johann Fischer, 15
Johann Friedrich Wolff, 277
Johann Wilhelm Meigen, 14
Johann Wolfgang, 7
Johannes Winnertz, 30
John Abbot, 260
John Fleming, 15
John Henry Comstock, 33
Johnson, 174, 175
Joseph Alexandre Laboulbène, 268
Justin O. Schmidt, 273

K

Karel Branèik, 261
Karl Eckstein, 40, 264
KFDV, 259
Kurimoto Masayoshi, 12

L

Lajos Abafi, 260
Lepidoptera, 104, 216
Lepidopterology, 3
Locomotion, 107
Longinos Navás, 270
Louis Agassiz, 260
Louis Jurine, 11

M

M. Grassberger, 68
Malacostraca, 109
Mantodea, 104, 215
Mantophasmatodea, 215
McGavin, 176
Medical Entomology, 256
Medicolegal forensic entomology, 55
Megasecoptera, 214
Metamorphosis, 101
Müllerian Mimicry, 184
Multiparasitism, 224
Murray S. Blum, 261
Muscular System, 166
MVEV, 259
Myrmecology, 3

N

Neuroptera, 104
Norman Tindale, 275

O

Odonata, 160, 215
Orthoptera, 135, 215
Orthopterology, 3
Oryzaephilus Surinamensis, 79
Osbert Salvin, 273
Oswald Duda, 264
Otto Staudinger, 274

P

Palaeodictyoptera, 160, 214
Palaeodictyopteroidea, 135
Paleogene, 143
Panthalassa, 136
PETA, 219
Phasmatodea, 215
Phenotypic Plasticity, 190
Philip Crowley, 263
Philip James Barraud, 260
Phylogeny of Insects, 129
Physiology, 129
Pietro Stefanelli, 75, 274
Plecoptera, 215
Plodia Interpunctella, 80
PMI, 67
Protorthoptera, 215
Pterygota, 214

R

Ray F. Smith, 76
Reproductive System, 173
Rhyniognatha hirsti, 110
Rob Denell, 78
Robert John Tillyard, 275

S

Scarab beetles, 59
SCP, 120
SIT, 191, 196
Staphylinidae, 143
Stephen Richards, 78
Sterile Insect Technique, 194
Sterility, 129
Superpara itism, 224

T

Takashi Shirozu, 274
Taxonomic Specialization, 3
Theodor Becker, 261
Thomas Desvignes, 264
Thomas Marsham, 10
Thysanura, 112
Titanoptera, 215
Triassic, 138
Trichopterology, 3
Triplehorn, 174, 175, 176

V

Victor Motschulsky, 270
Vincent Allard, 260
Vincent Wigglesworth, 276

W

W.D. Hamilton, 179
W.H. Evans, 265
WHO, 255
William Buckler, 262
William de 'wis, 260
William Elford Leach, 12
William Francis, 37
William Harris Ashmead, 260
William Kirby, 12
William Morton Wheeler, 37
William Schaus, 273
Willis Blatchley, 261
Wolfgang Schwenke, 273
World War II, 248

Z

Zoraptera, 215